PHY 1110 - General Physics Laboratory – II

5th Edition

Gopalan Srinivasan

AND

Rao Bidthanapally

Cover image © Shutterstock.com

www.kendallhunt.com
Send all inquiries to:
4050 Westmark Drive
Dubuque, IA 52004-1840

Copyright © 2019 by Kendall Hunt Publishing Company

ISBN: 978-1-5249-9576-8

All rights reserved. No part of this publication may be reproduced,
stored in a retrieval system, or transmitted, in any form or by any means,
electronic, mechanical, photocopying, recording, or otherwise,
without the prior written permission of the copyright owner.

Published in the United States of America

GENERAL PHYSICS LABORATORY-II-Experiments

TABLE OF CONTENTS

page

I. INTRODUCTION

    1. Emergency Information 1

    2. Lab Safety 2

    3. Notes 3

    4. Introductory Comments
        a) Goals 3
        b) Co-requisite 3
        c) Laboratory Reports 4
        d) Check list for a Laboratory Report............................ 4
        e) Preliminary Observations..................................... 4
        f) Final Data 5
        g) Analysis and Conclusions..................................... 5
        h) Graphs 5
            Graphical Analysis Software 5
        i) Summary 7

    5. Error Calculation – useful formulae 8

II. **LABORATORY EXPERIMENTS**

    Experiments 1 and 2: Ohm's law and DC and AC circuits 10

        Experiment 1: Ohm's Law and DC Series Circuits 17

        Experiments 2: DC Parallel Circuits and AC circuits 23

    Experiment 3: Discharging and Charging a Capacitor 29

    Experiment 4: Force on a Current Carrying Wire in a Magnetic Field 43

    Experiments 5: Focal Length of a Lens and Image Formation 59

    Experiment 6: Polarization of Light 73

    Experiments 7 and 8: Diffraction and Interference of Light 87

        Experiment 7: Single slit diffraction 103

Experiment 8: Double slit diffraction 109

Experiments 9 and 10: Helium and Hydrogen Line Spectrum 115

Experiment 9: Helium Line spectra 123

Experiment 10: Hydrogen Line Spectra 127

Experiments 11 and 12: Nuclear Counting Statistics and Interaction of Radiation with Matter 135

Experiment 11: Nuclear counting statistics 141

Experiment 12: Attenuation of gamma particles 145

I. INTRODUCTION

1. Emergency Information

- If there is an emergency situation, call campus police at **248-370-3331** or dial **911**.

- When calling the police, describe the emergency situation or nature of help needed and the room where help should arrive. ***The laboratory classes are held in room 269 Hannah Hall of Science (HHS).***

- Closest accessible telephone would be a cellular phone that you or someone in your class has.

FIRE: When there is a fire or if the fire alarm is activated, exit the building.

- Fire extinguishers are located in the lab. Take a minute to note their locations.

A FIRST AID KIT is available in the lab.

Student safety and equipment care are important. Safety precautions are discussed in the manual for each of the experiments. Please pay attention to those instructions while reading the manual. Your instructor will go over them in the class. Be sure that you understand and follow them.

It is IMPORTANT that you are aware of the Emergency procedures pertaining to this lab when you work here.

CAMPUS POLICE:	**248-370-3331** OR 911
Physics Department: Office (Rm. 190 SEB)	248-370-3416

2. Lab Safety

In the physics lab, the main safety concerns are electrical and laser. Throughout the semester, your instructor will remind you of possible hazards as these hazards present themselves.

1. Do not modify or alter the laboratory equipment in any way unless the modification is directed by the instructor.
2. Do not force any of the equipment. If an excessive amount of force is necessary tell your instructor or TA.
3. Electrical components and various lab items can become very hot during use – caution is necessary. Do not touch anything that may be hot, including electric immersion heaters and electric bulbs.
4. Several **electrical** devices are utilized in the physics lab. Most devices are designed to be safe under normal conditions and designed to operate at safe voltages and currents. Do not use settings other than those specified by your instructor or TA.
5. When required to build your own circuits, often bare wires may be present. Always use caution when dealing with bare wires.
6. When operating electrical equipment, ensure that the lab countertops, equipment, and your hands are dry. Check all cords and plugs to be sure that they are in good working condition. Look for exposed, frayed, or broken wires. Do not grab wires to unplug equipment. Hold the plug to disconnect equipment from the socket.
7. Turn power supplies off when you are changing significant parts of a circuit or making adjustments in the circuit.
8. **Lasers** may be used in some labs. When first turning on the laser, make sure it is pointed towards a wall or book and never toward an individual. Laser light that shines directly into the eye can cause permanent damage to the retina. Remember you cannot see the beam, so make sure you know its path!
9. When you have finished your work, check that electric circuits are disconnected. Clean your work area at the end of each lab.
10. If you have a question about safety you are to direct it immediately to your instructor or TA.

(Sign and date below for your reference. Another identical copy will be distributed on the first day of classes. Sign the copy and give it back to the instructors)

I agree that I have read and understand the Oakland University Undergraduate Physics Laboratory Safety Manual and will abide by the regulations of the Department of Physics at Oakland University as given in the Oakland University Safety Manual. Failure to abide by these rules and any conduct deemed unsafe by the instructor is intolerable and is grounds for immediate dismissal from the laboratory.

Signature: _____ Date: _____

3. Notes

Attendance in all class meetings is required. During the semester you are required to perform the experiments in the class for which you enrolled. There will NOT be any make-up classes for this course. In case of illness, it may be possible to do the missed lab in another section. Contact your TA or the professor in-charge for additional information. Grades for missed classes, reports and quizzes will be taken as zero.

Bring to all class meetings: Lab manual, scientific calculator, pencils, pens, erasers, and a clear plastic metric ruler.

Before you come to the laboratory session, read the experiment description in the manual and go through the relevant background material from your Introductory Physics text-book, if needed.

Before coming to the class, write out a one-page introduction for each experiment in the space provided in the manual. It should be a brief description, outlining in your words the objectives and procedures of the experiment.

During the class, perform the experiment and data analysis following the instructions in the manual and help from your instructor.

Handle all equipment carefully. If you damage anything (e.g. lens or power supply or multimeter, etc.), report it to your instructor.

A report, consisting of the introduction, data, graphs, calculations, analysis, and conclusions or summary, is due by the deadline specified in the syllabus.

> **The following are academic misconduct, leading to a grade of F for the course and referral to the Academic Conduct Committee:**
> - **Identical lab reports (data and graphs could be the same as partner's report)**
> - **Borrowed data or reports from previous years.**

Your reports will be graded based on your preparation, performance, quality and accuracy of data, analysis, and conclusions.

There will be 2 quizzes during the semester. The day and time for the quizzes are in the syllabus.

4. Introductory comments
a) Goals
In this course, you will:
- learn how to use basic physical measuring devices;
- become familiar with selected physical laws and phenomena;
- get experience taking data and drawing conclusions from them;
- learn how to estimate and to combine experimental errors.

b) Co-requisite
Co-requisite: PHY 1020 or 1520. The experiments deal with electricity, light, modern physics, and nuclear physics. It is very important that you STUDY THE MATERIAL IN THE TEXT even though you may not have reached it in the course. You will find the laboratory experience a great help to you in your subsequent course work.

The THEORY and EQUIPMENT sections of your experiments are covered in the laboratory manual. It is crucial that you become familiar with the information about each experiment in the laboratory manual and your text **before** coming to class. This way you will avoid wasting your valuable time in the class figuring out what should be done. Often times the equipment you will use are very delicate even though they may not appear sophisticated. Use adequate care and caution while handling them.

c) Laboratory Reports

Although you do the experiments with a partner, individual reports are required. Your reports will consist of INTRODUCTION (no more than one page), PRELIMINARY DATA, FINAL DATA, ANALYSIS, and CONCLUSIONS. You will use the necessary detachable pages from the manual. You should finish the experiment within the laboratory period. This will be possible only if you study the manual carefully ahead of time and come to the class with a plan for the experiment and report.

Your original data pages should be reasonably clear, but they will not be graded for neatness. You are encouraged NOT to recopy your original data pages. If you choose to do so you must include the original data pages with the instructor's initial on it as part of your report. Your **data must be recorded** in ink with no erasures. Incorrect entries may be corrected, or repeated but NOT ERASED. On the other hand, your calculations and graphs are better done in pencil so that you can make corrections and revise fits easily. ORIGINAL DATA PAGES MUST BE INITIALED BY THE INSTRUCTOR BEFORE YOU LEAVE THE CLASSROOM. It is your responsibility to get the data pages initialed by the instructor. Reports with uninitialed data pages will not be graded and be counted as zero. Please return your table to its original condition AFTER your instructor has checked your work and initialed the data pages.

Reports are due at the end of every two experiments.
The reports are due at the time and day indicated in your syllabus. You may drop the reports before the deadline in the boxes kept in front of your laboratory room.
Your report grade will be based on your preparation, performance, quality of data, analysis, and conclusions. Late reports will NOT be graded.

d) Check List for a Laboratory Report
- Is the Introduction brief and complete? Are all the steps presented clearly and the pages for additional work such as graphs, numbered in sequence?
- Are series of data with a varied parameter recorded in columns, or rows? Are the related fixed parameters also recorded?
- Are the units for every measured and calculated quantity shown correctly at all places?
- Have you recorded directly observed numbers, rather than differences done in your head? For example, if you measure the width of a table with a meter stick, it is best to work between two marks, say 10.0 cm and 93.2 cm, rather than using the battered end of the meter stick. But you should record the two positions, 10.0 cm and 93.2 cm, as well as the calculated length of 83.2 cm.
- Were the instruments read as precisely as possible?
- Are sample calculations shown?
- Have you taken all the required data, answered all questions in the manual, and recorded your conclusions as summary.
- Are the data sheets initialed by the instructor?

e) Preliminary Observations
- It is quite all right to make preliminary, <u>unrecorded</u> trials in order to get familiar with the apparatus.
- Record all data (initial and final) in ink, with no erasures. Mistakes may be crossed out. The reason for the revised numbers should be noted.
- Calculate the expected sizes of quantities whenever possible.

- Try different procedures, observers, and/or equipment in order to assess and reduce systematic errors.
- Take repeated runs and note smallest scale divisions in order to assess random errors.
- Find the maximum practical ranges of parameters in order to plan your final data.
- Make preliminary calculations and, where appropriate, graphs in order to see if your data is reasonable.

f) Final Data
- Final data should usually be graphed so that you check if data points are reasonably spaced and that their range is sufficient. Take additional data if required.
- Some data points should be repeated in order to see if there are equipment drifts or recording mistakes.

g) Analysis and Conclusions
- Sample calculations of derived quantities and errors should be included.
- Significant figures should be used properly.
- Final results should be displayed in tables and/or graphs whenever possible.
- Conclusions should be quantitative and specific and be based on measured and calculated numbers wherever possible. An exception: "How might you have improved the experiment using comparable equipment?" is admittedly speculative.

h) Graphs
- Each graph should have a TITLE.
- ERROR bars should be included if appropriate.
- The AXES should be labeled with the quantity plotted and its units.
- The scale for axes should be chosen so that the graph is approximately square and as large as possible; the positions of the data points are easy to read.
- When the slope of a straight-line fit is measured, two <u>widely separated points</u> (x_1, y_1), (x_2, y_2) on the line WHICH ARE NOT DATA POINTS should be used to evaluate the slope $S = (y_2 - y_1) / (x_2 - x_1)$. These points should be shown on the graph.

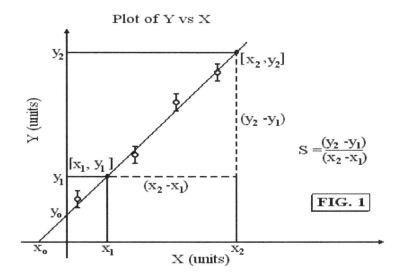

Using Computer and Graphical Analysis Software (Logger Pro 3 ver 3.8.6.1)

The computer and Logger Pro software are user friendly and it is very easy to learn to use them. The following set of instructions should help you getting started.

Switch on the computer the monitor. The computer should boot up. Then locate and double-click on Logger Pro icon. If you have difficulty starting the software, please consult your instructor. The Quick Reference Card below gives some information on getting started with the Program and some of its features.

IT IS IMPORTANT that every page you print out contains the title for the graph and the names of the persons printing it (you and your partners) in the text window or under print options for identification. Instructors will NOT grade graphs without printed identification from the computer.

Logger Pro
Quick Reference Card

To get started...
- Double-click on Logger Pro icon.
- When the program starts, you have a blank data set on the screen. Type in your X-value, press <return> and enter Y. Press <return> to go on to the next entry for (X, Y). The data will be graphed automatically as it is entered with X- values along the horizontal axis and Y-values along vertical axis.
To Enter a Title: Double-click on the graph (right side of screen) to enter title, axes labels, and other, under Graph/axis options.
To change the style of the graph, double-click on the graph window (right side of the screen) and then choose the style you want (bar graph, linear or Log or semi-log; ….. etc) under graph options.

- To change the scale of the graph: double click on the graph, the axes and type in a new value. You can also double-click on the axis of a graph to change the scale.
- To change what is graphed on an axis, click on the axis label. A pop-up menu will appear listing all the possibilities.
- To try a curve fit, choose Analyze from pull down menu. Select the general type of function you want to try. If you choose Manual curve fit, you can adjust the parameters and see how the fitted line and the mean square error change.
- To add a new column choose "new column" under "data" in the main screen.
- For error bars: Double-click on X or Y in the "data set." Choose "options" , then "error bar calculation" and "5%" for 5% error, then "done." The graphs will show horizontal bars that are 5% of x-values. Do the same for Y-error bars.

<u>Key features of Logger Pro</u>

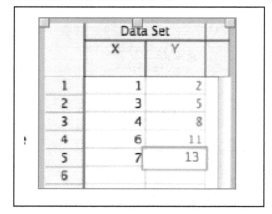

Typical data set entry (left side of screen)

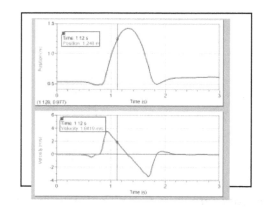

Typical graph (right side of screen)

Click	To	Click	To
▯	Create a new document.		Enter Tangent Line mode.
	Open an existing document.		Integrate selected data range.
	Save the active document or template with its current name.	STAT	Calculate statistics on selected data.
	Print the screen as displayed on your monitor.		Make a linear fit to the selected range.
	Scroll between pages.		Perform general curve fits.
	Bring up the previous page of the current document.		Set data-collection parameters.
	Bring up the next page of the current document.		Begin collecting data.
	Bring up the Data Browser Window.		*(The following icons only appear when collecting data or if relevant.)*
	Import data from the Texas Instruments calculator.		Stops collecting data or stops a replay.
A	Automatically scale the graph to include all data points.		Tells the heat pulser to generate a pulse.
	Zoom in to the selected region of the graph.		Click this button to keep a data point.
	Zoom out by a factor of two in both x- and y-directions.		Click the Zero button to set the current value of a sensor to zero.
	Enter Examine mode.		Create a new Graph Match.

- Student Versions of Graphical Analysis software may be purchased from
  Vernier Software
  8565 S.W. Beaverton-Hillsdale Hwy
  Portland, OR 97225-2429.
  Tel: (503) 297-5317; e-mail: dvernier@vernier.com

i) **Summary**
   - *Read the lab manual before coming to the class and come prepared with an introduction for the report and a plan regarding how to do the experiment.*
   - *Follow all steps outlined in the manual during the class.*
   - *Record data in INK only; no pencils. Do not overwrite; strike erroneous data and rewrite.*
   - *Write down the units of all measured and calculated quantities.*
   - *Present data legibly in tables and graphs.*
   - *Quick check the reasonableness of your data before taking a large number of data points.*
   - *Show sample calculations and present results in neat tables and graphs.*
   - *Summarize your conclusions based on the data and logical reasoning.*
   - *Turn-in the report in the drop-box for your class before the time specified in the syllabus.*

Error Calculation – useful formulae

If speed of sound is expressed as v ± Δv = 329 ± 10 m/s then the measurement has
ABSOLUTE ERROR, Δv = ± 10 m/s.
The FRACTIONAL (or RELATIVE) ERROR, Δv /v = ± 10/329 = ± 0.03; and
the PERCENT ERROR in v is Δv /v × 100 = ± 3%.
Percent difference [v(measured) – v(expected)]/ v(expected)x100%

Random errors

Random errors are easily found by looking at the scatter of repeated measurements. The random error Δx(random) is half the extreme variation [x(max) – x(min)], i.e.,

$$\Delta x(random) = \frac{1}{2}[x(max) - x(min)] \qquad \text{Random Error} \qquad (1)$$

For example, if we drop a meter stick four times and catch it after it has fallen the distances d = 16, 24, 22, and 20 cm, Δd = (1/2) (24 cm - 16 cm) = 4 cm. Δd = 4 cm is the uncertainty in every SINGLE TRIAL you made. Since we have gone to the trouble of taking four trials, the AVERAGE of these will be better information than a single trial.

$$x(av) = \frac{1}{N}(x_1 + x_2 + ...x_N) = \frac{1}{N}\sum_{i=1}^{N} x_i \qquad \text{Average} \qquad (2)$$

where N = the number of trials. For the above data, d(av) = (1/4) (16 + 24 + 22 + 20) = 20.5cm. The average is more precise than a single trial by a factor of $(1/N)^{1/2}$. That is, the uncertainty of the average is smaller than the uncertainty of a single trial.

$$\Delta x(av) = \frac{\Delta x(random)}{\sqrt{N}} = \frac{1}{2\sqrt{N}}[x(max) - x(min)] \qquad \text{Error in average} \qquad (3)$$

For the above data, Δd(av) = 4 cm/$\sqrt{4}$ = 4 cm/2 = 2 cm. Hence our best information about d is d(av) ± Δd(av) = (20.5 ± 2) cm = (21 ± 2) cm, when reported to be consistent with its error.

Combining Errors

When you are calculating a quantity, f, which is a function of several measured variables, x, y, z,.. with errors Δx, Δy, Δz, . . . , the rules for combining errors are as follows.

(1) A quantity MULTIPLIED BY A CONSTANT; if f(x) = Cx, then Δf = CΔx.
RULE: Multiply the error Δx by the constant C.

(2) ADDITION OR SUBTRACTION; if f (x,y) = (x + y) or (x - y), then
$$\Delta f = [(\Delta x)^2 + (\Delta y)^2]^{1/2}$$
RULE: take the square root of the sum of the squares of the errors Δx and Δy.

(3) A quantity RAISED TO A POWER n; if f(x) = Cx^n, then Δf/f = n(Δx/x).
RULE: take n times the <u>fractional</u> (or percent) error in x.

(4) MULTIPLICATION OR DIVISION; if $f(x,y,z) = C(xy/z)$, then

$$\Delta f/f = [(\Delta x/x)^2 + (\Delta y/y)^2 + (\Delta z/z)^2]^{1/2}.$$

RULE: take the square root of the sum of the squares of the *fractional* errors, or percent errors.

NOTE: In the above set of rules (1) and (2) are for absolute error Δf, and rules (3) and (4) are for relative error $\Delta f/f$.

EXAMPLES of the Rules for Combining Errors.

Rule (1) The time t for 20 oscillations of a mass on a spring is 15.4 ± 0.2 s.
The period, which is the time for one oscillation is,
$T = t/N = (15.4)/20 = 0.77$ s.
Error in T, $\Delta T = 0.2/20 = 0.01$ s. Therefore, $T \pm \Delta T = 0.770 \pm 0.010$ s.

Rule (2) Suppose $m_1 = 10.0 \pm 1.0$ g, $m_2 = (200.0 \pm 1.0)$ g and $M = m_1 + m_2$
$M \pm \Delta M = (10.0 + 200.0) \pm [1.0^2 + 1.0^2]^{1/2} = 210.0 \pm 1.4$ g.

Rule (3) Suppose $t = 3.40 \pm 0.10$ s, what is the error in $f = t^2$?
$f = t^2 = 11.56$ s^2. $\Delta f/f = 2(\Delta t/t) = 2(0.1/3.4) = 0.059$ no units.
Therefore, $\Delta f = f(\Delta f/f) = (11.56)(0.059) = 0.68$ s^2.
Rounding off: $f \pm \Delta f = 11.6 \pm 0.7$ s^2

Rule (4) Suppose $t = 3.40 \pm 0.10$ s, $g = 9.81 \pm 0.05$ m/s^2, and $h = (1/2) gt^2$
$h = (1/2) gt^2 = (1/2)(9.81)(11.56) = 56.7$ m.
The fractional error in g is $(0.05/9.81) = 0.005$.
The fractional error in t^2 is $2(0.1/3.4) = 0.059$ as shown above in Rule(3) example.
The factor (1/2) has no error.
Therefore, $\Delta h/h = [(0.005)^2 + (0.059)^2]^{1/2} = 0.0592 = 0.059$ (rounding off).
Then, $\Delta h = h (\Delta h/h) = 56.7$ m x $0.059 = 3.3$ m, and so $h \pm \Delta h = 57 \pm 3$ m.

Note: In this example, the fractional error in g may be neglected since it is ~ 10 times smaller than that in t^2. Therefore, the fractional error in h is equal to the fractional error in t^2.

Experiments 1 and 2: Ohm's Law and DC and AC Electric Circuits

Ohm's Law and DC circuits

Ohm's law states that the electric current, I, through an object will increase linearly with the voltage V applied across the object.

$$V = R\,I \qquad \text{(volts = ohms} \times \text{amperes)} \text{ (Ohm's Law)} \qquad (A)$$

The law is approximate. The resistance R, which is the constant of proportionality, is often only constant over limited ranges of current.

An ammeter, AM, is a low resistance device which is put in series with the circuit element whose current is to be measured. A voltmeter, VM, is a high resistance device put in parallel with the circuit element (s) whose voltage is to be measured. A multimeter has the provisions for both VM and AM and so it can be used either as a VM or an AM. Fig. 1 shows a DC source, R and VM and an AM connected to measure the voltage across R and current through it. The AM is reading both the current through R and that through the VM. Since the VM current is much smaller than the resistor current, we may take the AM current to be equal to the resistor current.

The resistance of our wire-wound resistors is accurate to 0.5% and constant to better than 0.01% over the range of currents we will use in this experiment.

Fig. 2 is a sketch of the circuit elements and wires connecting them. Until you get more experience wiring circuits, you should always draw both the circuit diagram AND a sketch of the circuit elements and wires.

Circuit Diagram   **Sketch of Circuit elements and wires**

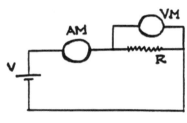 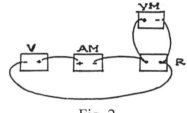

Fig. 1   Fig. 2

The power delivered by a voltage source, or the power dissipated by a circuit element is

$$P = I\,V. \qquad \text{(watts = amperes} \times \text{volts)} \qquad (B)$$

When power is dissipated by a resistor, equations (A) and (B) can be combined to give

$$P = I\,V = I^2 R \qquad P = V^2/R. \qquad (C)$$

One should always calculate the expected power dissipation in a resistor to make sure that its power rating is not exceeded.

Experiments 1 and 2: Ohm's Law and DC and AC Electric Circuits

DC Series Circuits:

Fig. 3 shows two resistors R_1 and R_2 connected in <u>series</u>. The total circuit resistance is

$$R_s = R_1 + R_2. \quad \text{(resistors in series)} \tag{D}$$

The voltages V_1 and V_2 across resistors R_1 and R_2 respectively, add to equal the supply voltage V_s across both resistors.

$$V_s = V_1 + V_2 \quad \text{(voltage addition)} \tag{E}$$

Thus series resistors are useful as voltage dividers.

Since the same current I flows through both resistors, the voltage drops V_1 and V_2 are proportional to the resistances, R_1 and R_2, i.e., $V_1 = I R_1$ and $V_2 = I R_2$. Therefore,

$$\frac{V_2}{V_1} = \frac{IR_2}{IR_1} = \frac{R_2}{R_1} \quad \text{(voltage division)} \tag{F}$$

Circuit Diagram   **Sketch**

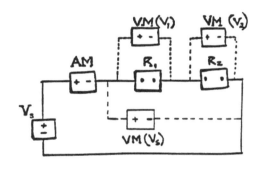

Fig. 3

Experiments 1 and 2: Ohm's Law and DC and AC Electric Circuits

DC Parallel Circuits:

Fig. 4 shows two resistors R_2 and R_3 connected in <u>parallel</u>. The conductances (1/R) of resistors in parallel add. i.e.,

$$\frac{1}{R_{23}} = \frac{1}{R_2} + \frac{1}{R_3} \quad \text{or} \quad R_{23} = \frac{R_2 R_3}{(R_2 + R_3)} \tag{G}$$

Fig. 4 also shows the parallel combination of R_2 and R_3 connected in series with R_3. The total resistance R in the circuit is $R_1 + R_{23}$. The total current through the circuit is I_1:

$$I_1 = \frac{V}{R} = \frac{V}{(R_1 + R_{23})} \tag{H}$$

At branch point (b), the current I_1 flowing through R_1 divides into currents I_2, through R_2, and I_3, through R_3.

$$I_1 = I_2 + I_3 \quad \text{(current addition)} \tag{I}$$

Equation (I) also holds at branch point (c), where I_2 and I_3 are the currents flowing into (c) and I_1 is the current leaving (c). Since the same voltage V_{23} is across both R_2 and R_3,

$$V_{23} = I_2 R_2 = I_3 R_3 \quad \text{and} \quad \frac{I_3}{I_2} = \frac{R_2}{R_3} \text{ (current division)} \tag{J}$$

Parallel resistors are useful as current dividers.

Circuit Diagram   **Sketch**

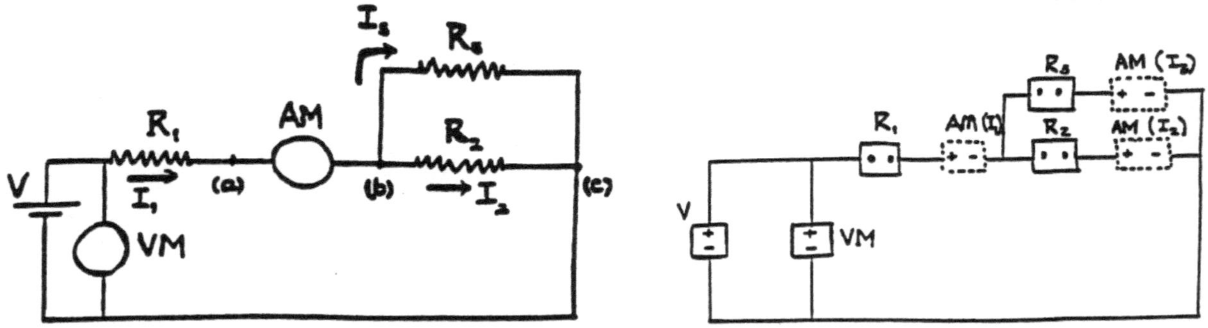

Fig. 4

Experiments 1 and 2: Ohm's Law and DC and AC Electric Circuits

AC circuits:

Figure 5 shows a circuit with R and an AC source with a voltage V that can be written as:

$V = V_p \cos(2\pi ft)$. Vp is the peak voltage and f is the frequency.

V could also be written as a sine function:

$V = V_p \sin(2\pi ft)$.

A circuit with an ac source and a resistance R is shown below. The ac voltmeter VM measures the voltage across R. Since the instantaneous voltage varies from a maximum of Vp to zero to -Vp, the meter measures the "root-mean-square (rms) voltage" V_{rms}.

V_{rms} is related to the peak voltage Vp by the relation: $V_{rms} = V_p/\sqrt{2} = 0.707\, V_p$.

The ammeter also measures the root-mean-square current $I_{rms} = I_p/\sqrt{2} = 0.707\, I_p$.

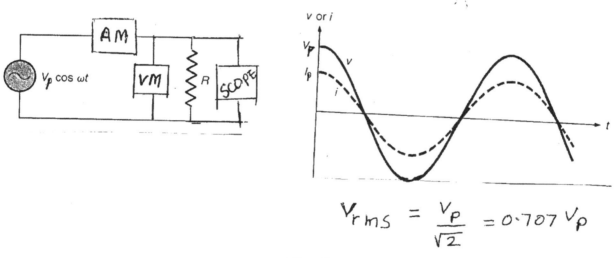

$$V_{rms} = \frac{V_p}{\sqrt{2}} = 0.707\, V_p$$

Fig. 5

Using an oscilloscope to measure ac voltages
You will use an oscilloscope to measure the ac voltage across a resistance.

Oscilloscopes
An oscilloscope is a device which displays any periodic voltage as a function of time. A light spot is moved repeatedly to the right at the same time it is deflected vertically by the voltage V. For slow voltage variations and low sweep speeds, one sees a moving spot. As the spot is swept faster and more frequently, one sees a repeatedly traced line because the light given off by each point on the line persists between the spot traversals.

We can use an oscilloscope to determine the period (and thus the frequency) and also the amplitude (voltage) of any periodic waveform. As mentioned before, we will work with an ac voltage (sinusoidal waveforms only). Distances along the x-axis are directly proportional to time and distances along y-axis are proportional to voltage. Thus, the y versus x picture displayed on the oscilloscope screen becomes a V versus t plot (Fig. 5).

An oscilloscope (Fig. 6) can be used to display waveforms as well as to measure the period or frequency (along horizontal axis) and the voltage (along vertical axis) of the displayed waveforms. In the experiment, you will use AC voltage source. The oscilloscope you will use is a DUAL CHANNEL OSCILLOSCOPE, which can display either or both the waveform input given to CH-1 and CH-2. The controls labeled under DISPLAY and the horizontal and vertical position control knobs are self-explanatory. The oscilloscope screen is divided into large divisions along the **x-axis** and along the **y-axis**. Each of the division (one square) is further divided into 5 parts so that measurements can be made precise to 0.2 division. **Knobs under "HORIZONTAL," "VERTICAL," and "TRIGGER" labels control the SWEEP RATE (time/div), GAIN (volt/div), and triggering of the waveform, respectively.** Since an oscilloscope is a general purpose device that can be used to measure voltages and times of various magnitudes, there are several sweep rate [Horizontal (time/div)] and gain [Vertical (volts/div)] settings. You should learn and understand the function of these controls. Knobs labeled "VARIABLE" (located to the right of sweep rate control and the inner knobs on Gain control for CH-1 and 2) provide an uncalibrated control. They should be switched off (turned completely counter clockwise) during your experiment.

Take some time to familiarize yourself with all the different knobs and their labels.

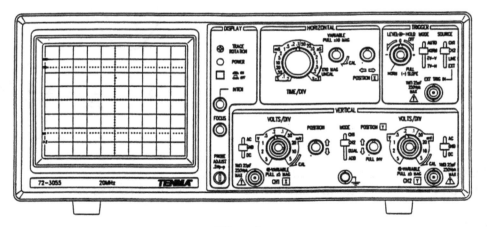

Fig. 6

The vertical scale on the oscilloscope screen can be used for voltage measurements. (Recall: the horizontal axis is time.) The gain control setting G (volts/div) simply converts vertical deflection to volts. The amplitude V of a voltage form is its center-to-peak voltage. In practice, it is easier to read the peak-to-peak voltage 2A (Fig. 7). This way the error involved in finding the mid-point of the waveform can be avoided in your measurements.

Suppose the waveform displayed on the oscilloscope screen is as shown in Fig. 7. Then the peak-to-peak voltage is given by,

$$2A(\text{volts}) = G(\text{volts/div})(y_2-y_1)(\text{div}),$$

where (y_2-y_1) is the peak-to-peak height measured on the oscilloscope screen.

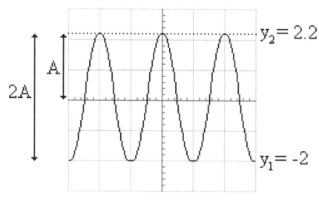

Fig.7

To read the y positions, adjust the vertical position so that the bottom of the signal falls on a horizontal line, such as $y_1 = -2$ cm (we assume that the origin is set at the center of the oscilloscope screen and each square on the screen corresponds to 1 cm length). Then adjust the horizontal position so a peak of the signal falls on the center vertical axis. Suppose the top of the signal falls at $y_2 = 2.2$ cm and the gain control is set at G = 0.1 V/div. Then,

$$2A = (0.1 \text{ V/cm})(2.2-(-2.0)) \text{ cm} = 0.1 \text{ V/cm} \times 4.2 \text{ cm} = 0.42 \text{ V}. \text{ Therefore, } A = 0.21 \text{ V}.$$

ELECTRICAL SAFETY RULES

I. Never work alone when your equipment has more than 25 volts on exposed terminals. (Voltages below 25 V are safe because the currents they can produce in the body are too small to cause physical damage. This experiment uses voltages less than 25 V, and therefore is safe.)

II. Always turn off voltage supplies and unplug them from the 110 V socket before touching, or re-wiring a circuit.

III. After doing II, touch the parts of the circuit lightly with the BACK OF THE FINGERS OF ONE HAND. Keep your other hand and feet away from any ground plates or pipes. (Reasons - If you get an electrical shock, your hand will automatically pull away from the circuit. You don't want another part of your body to complete a circuit.)

EQUIPMENT

1 – variable voltage AC/DC power supply (central scientific)
3 – decade resistance boxes (two 0-999 ohms and one 0-9999 ohms)
4 – digital multimeters
4 – medium, or long hook-up wires
4 – short, or medium hook-up wires
5 - Oscilloscope
For the room: Spare hook-up wires, resistance boxes, multimeters, fuses, screwdriver

Equipment Safety
1. Before switching on the power supply, make sure that <u>not</u> all resistance boxes are set to zero.
2. Turn off the power supply before changing R values. If the resistance selector switch is turned through zero with the voltage on, then a short circuit occurs and the power supply is damaged.
3. Do not exceed the voltage and current ratings for the resistances used.

BEFORE YOU BEGIN, the instructor will:
(a) review the ELECTRICAL SAFETY PRECAUTIONS;
(b) show you how to set the decade resistance boxes;
(c) show you the meter input terminals, range, and switch settings.

The OBJECTS of this experiment are:

I. to learn to wire series and parallel circuits;
II. to learn to insert voltmeters and ammeters properly in circuits;
III. to become familiar with voltage and current addition and division in circuits;
IV. to become familiar with ac circuits.

Experiment 1: Ohm's Law and Series DC Circuits: Introduction

Name: _____   Section/ Group:_____

Partner's name:_____   Instructor's Initial:_____

Experiment 1: Ohm's Law and Series DC Circuits

1-1: Preliminary Observations and Calculations:

(1) *Set the source on DC.* WIRE THE CIRCUIT of Figs. 1 and 2. Check your wiring by following the wires of Fig. 1 with one hand and those of the wired circuit with the other hand. Set R = 500 ohms. Set the voltage control to minimum V (full CCW - counterclockwise).

(2) RESISTOR POWER DISSIPATION. You are going to use R = 600 and 250 ohms. CALCULATE the maximum voltages and currents V(250), I(250), V(600), and I(600) which will give a power dissipation of 0.25 W for each R value. [Use Eq. (C).] Check your answers with the instructor. DO NOT EXCEED these V and I values.

Data

PLAN AND RECORD 10 to 12 data points, (V, I), for R = 250 ohms and five more for R = 600 ohms which are fairly evenly spaced between V(min) and V(max). Record data in units of volts and mA.

Graphs

PLOT I (along vertical axis) vs. V (horizontal axis) for <u>both</u> R values on one graph using Logger Pro. Fit straight lines to each set of I-V data and record their slopes along with their respective errors. SLOPE = 1/R. You have to include an assumed point at V = 0, I = 0.
RECORD extra data points if they are needed to establish the fits.

1-2: Series Circuits

The circuit diagram in Fig. 3 shows a voltmeter, VM, connected between points (a) and (c) so that it measures the voltage V_s. Use two other VM between points (a) and (b) and between (b) and (c) to measure V_1 and V_2.

Preliminary Observations

(1) SKETCH the components of the series circuit of Fig. 3 and wire the circuit. Set R_1 = R_2 = 150 ohms.

(2) CALCULATE the PREDICTED current I for V_s = 2.40 V. [Recall: $I = V_s / R_s$ where $R_s = R_1 + R_2$]

(3) MEASURE AND RECORD the current I. If it is within about 10% of the predicted value, continue. Move the voltmeter leads appropriately and MEASURE AND RECORD V_1 and V_2. If V_1 and V_2 are within about 5% of their expected values of 1.20 V, continue.

Data

We wish to look at voltage addition and division [Eqs. (E) and (F)] with a constant total series resistance $R_s = R_1 + R_2$ = 300 ohms and V_s = 2.4 V. Do not set V_1 or V_2 to equal to V_s). The current, I, will be held constant for all data.

(1) RECORD: I, R_2, R_1, V_2, V_1, and V_s for (R_1, R_2) values of: (150, 150), and five more combinations of R_1 and R_2, so that $R_1 + R_2$ = 300 ohms. For example, (120, 180), (90, 210), (60, 240), (30, 270) ohms.

(2) CALCULATE AND RECORD: ($V_2 + V_1$), V_2 / V_1, and R_2 / R_1.

Repeat any data points that appear questionable.

Experiment-1: Ohm's Law and Series DC Circuits   Data

Ohm's Law

1-1: Preliminary Observations and Calculations:

Power P = 0.25 W

Maximum Voltage(V) for R = 600 ohms: _____

Maximum Current(I) for R = 600 ohms: _____

Maximum Voltage(V) for R = 250 ohms: _____

Maximum Current(I) for R = 250 ohms: _____

Show V_{max} and I_{max} calculations:

Estimated voltage error ΔV: _____

(This is the reading error, know how to estimate the reading error)

Estimated current error ΔI: _____
(This is the reading error, know how to estimate the reading error)

Basis for estimates:

Experiment-1: Ohm's Law and Series DC Circuits -   Data (cont'd)

Data

R = 600 ohms

V	I

R = 250 ohms

V	I

Experiment-1: Ohm's Law and Series DC Circuits Data (cont'd)

1-2: Series Circuits

Preliminary Observations

Sketch:

When $V_s = 2.40$ V, calculate expected current I: _____
(Show calculations)

Calculated current I (for $V_s = 2.40$ V) _____   Observed current I _____

Percent difference _____

R_1	R_2	V_s	V_1	V_2	I	$V_1 + V_2$	V_2/V_1	R_2/R_1

Experiment-1: Ohm's Law and Series DC Circuits   Analysis

Ohm's law

(1) What do the slopes of the I vs V plots made in the Ohm's Law experiment represent?

(2) Find the resistance from the slopes of I vs. V graphs. Compare with the known R values. (Be careful about the units and convert them to base units if necessary)

(3) Find the percent difference between the R values calculated from the slope and known R values.

SERIES CIRCUIT

(4) Refer to the data in the table in Page 21: why the current through R_1 and R_2 must stay the same?

(5) For the data in the table in Page 21: Are the ratios V_2/V_1 and R_2/R_1 same? What is the maximum percentage deviation between V_2/V_1 and R_2/R_1?

Summary and Conclusions

Experiment 2: DC Parallel Circuits and AC Series Circuits- Introduction

Name:_____   Section/ Group:_____

Partner's name:_____   Instructor's Initial_____

Experiment 2: DC Parallel circuits and AC Series circuits

2-1: DC Parallel Circuits

Preliminary Observations:

Set up the circuit shown below: use 1 voltmeter, 3 ammeters, 3 resistance boxes and a DC power supply. Set $V_s = 7.00$ V, $R_1 = R_2 = R_3 = 200$;

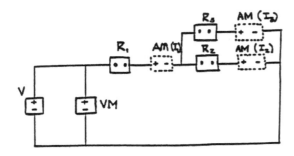

Calculate the expected current I_1: _____
(Show detailed calculations) _____

Calculated current $I_1 = $ _____   Observed current $I_1 = $ _____

Percent difference: _____

Data

Record: the currents I_1, I_2, I_3 for the R-values in the table. Record currents in mA.

Calculate and record: (I_2+I_3), I_3/I_2, and R_2/R_3.

Compare the values in the last 2 columns. Are they equal?

$R_2(\Omega)$	$R_3(\Omega)$	I_1	I_2	I_3	V_s	I_2+I_3	I_3/I_2	R_2/R_3
200	200							
250	167							
333	143							
500	125							
1000	111							

2-2: AC series circuits

Set up the circuit shown below.

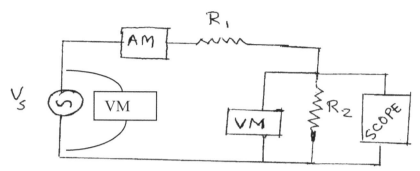

Switch AM and VM to ac mode. Use AC power supply. Set $V_s = 3$ V by measuring it with the VM across the source. Set $R_2 = 1000$ Ohms. R_1 values are given in the table below.

Measure the rms current I (rms) through R_1 and R_2. Measure the rms voltage V_2 (rms) across R_2. Measure the peak voltage $V_{2,peak}$ across R_2 with the oscilloscope. **Keep Vs = 3.0 V all the time.**

Compare the values $\sqrt{2}\ V_2$ (rms) and V_{2peak} (the last two columns in the table below).
Plot I (rms) versus V_2 (rms). Fit it to a linear curve and find the slope, S. Calculate $R_2 = 1/S$.

$R_2(\Omega)$	$R_1(\Omega)$	I (rms)	V_2 (rms)	$\sqrt{2}\ V_2$ (rms)	$V_{2,peak}$
1000	300				
1000	400				
1000	500				
1000	600				
1000	700				
1000	800				
1000	900				
1000	1000				
1000	1100				
1000	1200				

Experiment 2: DC Parallel circuits and AC Series circuits - Analysis

Parallel Circuit

(1) In a parallel circuit, does the current through the resistors or voltage across the parallel resistors ramain the same? Support your answer with numbers from your data and Equ. J.

(2) Using Equ. I, compare your observed values to the expected total current I_1. Find the percentage deviation of the "worst case' value from the expected value.

AC circuit:

(3) Compare the values $\sqrt{2}\ V_2$ and $V_{2,peak}$. Are they equal?

(4) The AC circuit used in this exercise is called a voltage divider. Why?

Summary and Conclusions

Experiments 1 and 2: Review Questions

(1) What are the objects of the experiment?

(2) (a) What are the electrical safety rules?
(b) Below what voltage is electricity safe?

(3) In a plot of V versus I, which resistor will give a straight line with the steeper slope, 200 ohms, or 500 ohms?

(4) What are the maximum voltage and current which should occur
(a) for a (1/4) watt, 300 ohm resistor? (b) for a (1/4) watt 3000 ohm resistor?

(5) (a) Re-draw Fig. 3, showing where to put the voltmeter to measure V_S.
(b) Re-draw Fig. 4, showing where to put the ammeter to measure I_2.

(6) Answer VOLTMETER, or AMMETER:
(a) Has a low resistance;
(b) Has a high resistance;
(c) Is connected in parallel with a circuit element, or elements;
(d) The circuit must be rewired in order to include it;
(e) It can be added to the circuit without rewiring the circuit.

(7) (a) For a given power dissipation, does V(max) INCREASE, or DECREASE with increasing R?
(b) Does I(max) INCREASE, or DECREASE with increasing R?

(8) Given V = 10 V in series with R_1 = 100 ohms and R_2 = 400 ohms.
(a) Calculate $I = V/(R_1 + R_2)$.
(b) Calculate V_1 and V_2.
(c) Calculate the power dissipated in each resistor.

(9) (a) Calculate I and V_2 in Fig. 3 when V = 6.00 V, R_1 = 200 ohms, and R_2 = 200 ohms.
(b) Calculate I_1, V_{23}, I_2, and I_3 for the circuit of Fig. 4 when R_3 = 300 ohms is added in parallel with R_2. Use same values for other quantities given in (a).
(c) Does the circuit current I_1 INCREASE, OR DECREASE, when R_3 is added? Explain why qualitatively.
(d) Does the voltage across R_2 INCREASE, OR DECREASE, when R_3 is added? Explain why qualitatively.

(10) (a) What is the expected slope of the plot in which the current I is along the horizontal axis and the voltage V is plotted on the vertical axis?
(b) What is the expected slope of the plot in which the current I is along the vertical axis and the voltage V is plotted on the horizontal axis?

(11) In an AC circuit with V_S = 10 V and series resistances R_1 = 100 Ω and R_2 = 250 Ω, find the rms current I, rms voltages V_1 and V_2 and peak voltages $V_{1,peak}$ and $V_{2,peak}$.

Experiment 3: Discharging and Charging a Capacitor

(i) Capacitor Discharge

Capacitors are charge storing devices. In an electrical/electronic circuit, capacitors can be used to store/release charge and thereby controlling the current. Fig. 1 shows a resistor R_1 connected across a capacitor C. Their common voltage is read by a voltmeter V having an internal resistance R_v. The resistance R_s (100 ohms) is connected in series with the power supply to protect it, when charging the capacitor. The capacitor is first connected to a **DC power supply** and charged to an initial voltage V_o. When the power supply lead is disconnected, the capacitor C will begin to discharge through R_1 and R_v in parallel. The voltage V will decrease exponentially with time.

$$V = V_o e^{-t/\tau} \tag{A}$$

where $\tau = RC$ (characteristic decay time constant)

Tau, τ, is the characteristic decay time constant in which the voltage falls from V_o to $(1/e) V_o$ (Fig. 2). Where e (= 2.718) is the base of natural logarithms. R is the parallel resistance of R_1 and R_v, the voltmeter resistance.

$$\frac{1}{R} = \frac{1}{R_1} + \frac{1}{R_v} \quad \text{or} \quad R = R_1 R_v / (R_1 + R_v) \tag{B}$$

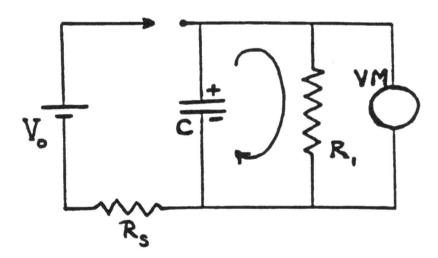

Fig. 1

Experiments 3   (i) Capacitor Discharge

Equation (A) may be rewritten using base 2, rather than base e.

$$V = V_0 2^{-t/T} \tag{C}$$

where $\quad T = 0.693\tau = 0.693RC \quad$ (half-life)

T is the HALF-LIFE of the decay, the time in which the voltage falls from V_0 to $(1/2) V_0$. V will fall by a factor of $(1/2)$ over the time interval T. Therefore over time intervals $t = 1T, 2T, 3T, 4T...$, V will decrease by $(1/2)^1, (1/2)^2, (1/2)^3, (1/2)^4 \ldots$ Thus

$$V_n = (1/2)^n V_0 \tag{D}$$

when $t = nT$.

Therefore, if we take V versus t data like that of Fig. 2, and draw a best-fit curve, the voltages $V_0, (1/2) V_0, (1/4) V_0, (1/8) V_0, (1/16) V_0 \ldots$ should occur at equally spaced time intervals T.

Knowing T and R, we can then calculate the capacitance C using

$$T = 0.693 \, RC \tag{E}$$

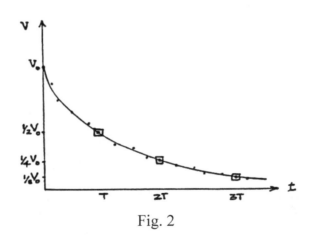

Fig. 2

It is always easiest to DRAW a straight-line best-fit curve. A plot of lnV versus t should be a straight line with a slope $-(1/\tau) = -(1/RC)$. This can be seen by taking the natural log of both sides of equation (A).

$$\ln V = \ln(V_0 e^{-t/\tau}) = -(1/\tau) \cdot t + \ln V_0 \tag{F}$$

Since the vertical axis of semi-log graph paper has a logarithmic scale, a plot of V versus t on semi-log paper should also be a straight line with a negative slope. Hence a semi-log plot of V versus t will tell us whether V does indeed decay exponentially. From the slope of the best-fit straight line we can determine τ and C as explained above.

Experiments 3   (ii) Capacitor Charge

(ii) Capacitor Charge

When a capacitor is charged from $V = 0$ through a resistor R_1 to a final voltage V_0 using the circuit shown in Fig. 3, the voltage increases exponentially in time with the same characteristic time, $\tau = RC$, and half-life, $T = 0.693\, RC$, as it does when the capacitor discharges. The charging equation is:

$$V = V_0(1 - e^{-t/\tau}) = V_0(1 - 2^{-t/T}) \qquad (G)$$

Note that equation (G) gives $V = 0$ for $t = 0$ and $V = V_0$ for $t =$ infinity. The difference between V and the final voltage V_0 decreases as $(1/2)^n$, and therefore $V = (1/2)\,V_0$, $(3/4)\,V_0$, $(7/8)\,V_0$, and $(15/16)\,V_0$ for $t = 1T, 2T, 3T$ and $4T$.

If we plot capacitor charging voltage V versus t and draw a best-fit curve, we can locate the $(1/2)\,V_0$, $(3/4)\,V_0$, $(7/8)\,V_0$, and $(15/16)\,V_0$ points and see if they are equally spaced in time in order to find out if the charging is exponential. We can then see if the charging half-life is the same as the discharging half life for the same resistance R_1.

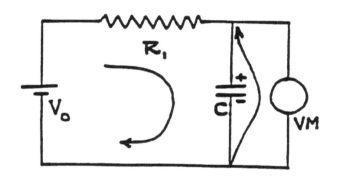

Fig. 3   Fig. 4

EQUIPMENT
1 - electrolytic capacitor of about 0.025 farads; 2 - resistance boxes; 1 - voltmeter
1 - DC voltage source, 2 to 10 volts, approx.
2 - interval timers
7 - hook-up wires
   extra hook-up wires (for the room)

Experiment 3: (i) Capacitor Discharge   Procedure

The OBJECTS of the experiment are:
I. To see if capacitor discharge is exponential in time;
II. to see if the half-life T varies directly with R;
IV. To determine C from the discharge data, given R_1, and R_v.

BEFORE YOU BEGIN, the instructor will remind you of the following equipment precautions.
- I. Turn off the voltage supply before wiring, or rewiring the circuit.
- II. Use the TAP TEST when you first apply voltage to a circuit.
- III. The CAPACITOR IS POLARIZED. Be sure to connect its positive terminal to positive voltage terminal and negative terminal to the negative terminal of the voltage supply.

Preliminary Data
(1) RECORD:
(a) The make and model of your voltmeter;
(b) The full-scale voltage you will use;
(c) The voltmeter resistance, R_v at the scale setting you will use.
(d) The smallest scale division in volts;

(2) WIRE the capacitor discharging circuit of Fig. 1. SKETCH the components and wires.

(3) Set R_s = 100 ohms and R_1 = 1000 ohms. Adjust the supply voltage to 2.8 V(3 volt scale). Disconnect the power supply lead and observe the discharge of the capacitor. RECORD: V_o and the time it takes the voltage to drop from V_o to $(1/2) V_o$ and to $(1/4) V_o$, after disconnecting the lead at the power supply. COMPARE the observed half-life with the calculated half-life T = 0.693 RC. If your observed and calculated values agree to about 20%, your circuit, procedure, and calculation are probably satisfactory.

(4) You are to take two sets of V vs. t capacitor discharge data similar to those shown in Fig. 2 for two values of R_1 varied as widely as possible. For each run, use the lap timer to record time at pre-determined voltage intervals. Practice taking the preliminary data at least couple of times before taking the final data. Your runs should last at least three half-lives and have more than 10 data points for each R_1. RECORD some preliminary data for one R_1 value. DON'T FORGET TO **RESET** THE TIMER EVERYTIME BEFORE TAKING DATA.

Final Data
(5) RECORD FINAL CAPACITOR DISCHARGE runs for two different values of R_1. RECORD estimated reading errors for voltage and time and discuss briefly the basis for the estimates.

Experiment 3: (ii) Capacitor Charge — Procedure

Capacitor Charging Procedure:

Preliminary Data

(1) WIRE the capacitor charging circuit of Fig. 3. SKETCH the components and wires. Don't forget the capacitor shorting wire, connected only to the negative terminal of the capacitor.

Before the capacitor is charged, its voltage is reduced to zero by temporarily connecting (shorting) a wire across its terminals. See Fig. 3. R_1 is then connected to V_o, after the shorting wire has been disconnected, to start the charging cycle.

Practice the following charging procedure.
 (a) Set $R_1 = 10$ ohms and connect it to the voltage supply. Set $V_o = 2.8$ V (A small value of R_1 is used in order that the capacitor voltage responds quickly to changes in the supply voltage. Try $R_1 = 1000$ ohms one time in order to see why the smaller value of R_1 is desirable when you are setting V_o.)
 (b) Disconnect R_1 from the voltage supply and short capacitor terminals several times, using the shorting wire, in order to be sure its voltage is 0.
 (c) Change R_1 to the value wanted for the charging data run.
 (d) Simultaneously connect R_1 to the voltage supply and start the timer.
 (e) Read t and V points as the capacitor charges using the lap timer.

(2) For $R_1 = 1000$ ohms. RECORD: V_o; the times for the capacitor to charge to $(1/2) V_o$ and to $(3/4) V_o$. COMPARE the observed charging half-life with the discharging half-life T found in **discharging experiment**. If the two are within about 10% of each other, your circuit and procedure are probably satisfactory.

Final Data

You are to take capacitor-charging data similar to those of Fig. 4 for the same two values of R_1 used in step (4) of the discharging experiment. Again, you should take at least 10 data points over at least three half-lives for each R_1.

RECORD final capacitor charging runs for two values of R_1 used in the discharging data.

Experiment 3: Capacitor Discharge and Charge - Graphs

Graphs

Using the computer:

(1) Make <u>two</u> graphs for the discharging and discharging data, one for each values of R_1 chosen for your final data. Plot V (vertical axis) versus I (horizontal axis). Fit the curve to the natural exponent.

(2) Make one ln V vs. t graph for both discharging runs. (This plot is equivalent to plotting V vs t on a semi-log graph paper. Fit it to a linear function and obtain the slope ($=1/\tau$) and error slope.

(3) Make a plot of **ln (1-V/V$_0$) vs t** for both charging runs. Do a linear fit and find the slope and error slope. Note $\tau = RC = 1/|\text{slope}|$.

Experiment 3: Capacitors Discharge and Charge — Introduction

Name: _____ **Section/Group:**

Lab Partners: **Instructor's initials:**

Introduction:

Experiment 3. Capacitor Discharge — Data

Capacitor Discharge - Preliminary Data

Sketch:

Full Scale voltage _____

Meter Resistance R_v _____

Smallest scale division _____

$V_o =$ _____

$R_1 =$ _____

$C =$ _____

t	V

Calculate the half-life time constant T:
(Show all calculations)

$T = 0.693 RC$
$R = R_1 R_v / (R_1 + R_v)$

Calculated T: _____

Observed time for $(1/2) V_o (=T)$ _____

Observed time for $(1/4) V_o (=2T)$ _____

Observed T: _____

% difference: _____

Experiment: 3 – Capacitor Discharge   Data (cont'd)

Capacitor Discharge - Final Data

Estimate for ΔV: _____   Basis for error estimates:

Estimate for the Δt:
(Assume a reasonable value for instrument error and also your reaction time, i.e., time it takes to note the time from the measuring device). _____
Basis for Δt error estimates:

$R_1 =$ _____ $R_1 =$ _____

t	V

t	V

In the Logger Pro create a calculated column to calculate ln(V) to plot ln(V) vs t.

Experiment 3: Capacitor Charge Data

Capacitor Charge - Preliminary Data

Sketch:

$V_o =$ _____ $R_1 =$ _____ $C =$ _____

t	V

Calculate the half-life time constant T:
(Show all calculations)

$T = 0.693 RC$
$R = R_1$

Calculated T: _____

Observed time for $(1/2) V_o (=T)$ _____

Observed time for $(3/4) V_o (=2T)$ _____

Observed T: _____

% difference: _____

Experiment 3: Capacitor Charge Data (cont'd)

Capacitor Charge - Final Data

Estimate for ΔV: _____   Basis for error estimates:

Estimate for Δt: _____   Basis for error estimates

$V_0 =$

$R_1 =$ _____ $R_1 =$ _____

t	V

t	V

In the Logger Pro create a calculated column to calculate $\ln(1-V/V_0)$ to plot $\ln(1-V/V_0)$ vs t.

Experiments 3: Capacitor Discharge and Charge - Analysis

(1) For each total parallel resistance R from discharge data, calculate the characteristic decay time τ, which is given by, $\tau = RC = 1/|\text{slope}|$.

(2) Calculate the value of capacitance, C, for each run given by: $C = \dfrac{\tau}{R} = \tau \cdot \left[\dfrac{1}{R_1} + \dfrac{1}{R_v} \right]$

(3) From the 2 estimated values of C find: C (average), ΔC (random), ΔC (average), and the percentage deviation of C(average) from the expected value of 25,000 µF.

(4) Are the charging and discharging half-lives (T) through the same resistance the same? Support your answer with numbers.

(5) Calculate C from the charging data using the slope of $\ln(1-V/V_0)$ vs t. Does your value agree with the number 25,000 µF label on the capacitor? Assume a manufacturer specified accuracy of ± 15%.

Summary and Conclusions

Experiment 3: Review Questions

(1) What are the objects of the experiment?

(2) Why should R_1 be set at a small value when adjusting V_o in the charging circuit.

(3) (a) For a discharging capacitor, the voltage initially at $V_o = 4.0$ volts drops to 0.5 volts in 90 s. What is the half-life for the capacitor discharge?

   (b) How long will it take the capacitor to charge from 0 to 3.0 volts if it is connected to a supply at 4.0 volts through the same resistance previously used in the discharge?

(4) (a) Given $C = 0.02$ F, calculate the resistance R which will give a 30 s half-life for charge or discharge.

   (b) Suppose that $R_v = 40,000$ ohms. What value of R_1 is needed to give a 30 s half-life?

(5) Use Equation (G) to show that $V = (7/8) V_o$ when $t = 3T$ in capacitor charging.

ANSWERS

(3) (a) 30 s   (b) 60 s

(4) (a) 2164 ohms   (b) 2288 ohms

Experiment 4: Force on a current carrying wire in a magnetic field

Force on a charged particle

A charged particle moving in a magnetic field with a velocity will experience a force. The magnitude of the force is given by

$$\mathbf{F} = q\mathbf{v} \times \mathbf{B} \; ; \qquad \text{magnitude of force } |\mathbf{F}| = qvB\sin\theta \qquad (A)$$

Where q is the charge on the particle, **v** is the velocity of the particle, **B** is the magnetic field and θ is the angle between the magnetic field and velocity. It is obvious from the expression that the force is maximum when velocity is perpendicular to the magnetic field (θ = 90°) and zero when velocity is parallel to the field (θ = 0°). The direction of the force is perpendicular to the plane containing velocity and magnetic field and is given by the right hand rule as shown in the fig. 1.

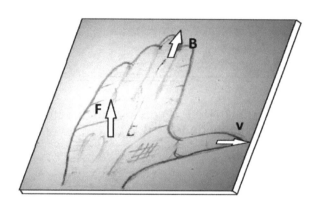

Fig.1: Right hand rule-1 (RHR-1) giving the direction of the force on a positive charge moving with a velocity **v** in a magnetic field **B**.

In Fig.1 when fingers point towards the magnetic field, thumb towards the velocity then the palm indicates the direction of the force on a positively charged particle. The equation A above can also be used to define the magnetic field.

$$\mathbf{B} = F/qv\sin\theta \qquad (B)$$

If a charge of 1C travelling at right angle to the magnetic field at a velocity of 1m/s and experiences a force of 1N then the uniform magnetic field in the region is one Tesla, T. One T = 1000 Gauss (Gauss, G is another unit for magnetic field).

Force on current carrying wire in a magnetic field

As the current is rate at which charges flow, one can expect a current carrying wire, placed in a magnetic field also experiences force due to the magnetic field. We can use the equation A to arrive at an expression to calculate the magnitude of the force on a current carrying wire placed in a magnetic field. Multiplying and dividing right hand side of the expression A by t and realizing current, I = q/t and length, L = v.t we get

$$\mathbf{F} = ILB\sin\theta \qquad (C)$$

Experiment 4: Force on a current carrying wire in a magnetic field

In the above expression, I is the current, L is the length of the wire in the magnetic field and **B** is the magnetic field strength. The θ is the angle between the magnetic field and current (or length of the wire as the current flows along the length of the wire). Once again the direction of the force can be found using the right hand rule; with the thumb pointing in the direction of conventional current, fingers pointing towards the magnetic field, the palm indicate the direction of the force on the wire. This is shown in figure 2.

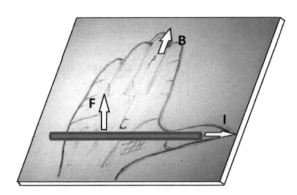

Fig.2: Right hand rule-1 (RHR-1) giving the direction of force on a current carrying wire, with a current I in a magnetic field **B**.

EQUIPMENT

- Basic Current Balance (PASCO model SF-8607)
- DC Power supply providing current up to 5A (PASCO model SF-9584)
- DC ammeter capable of measuring current up to 5A (A digital multi-meter can used)
- Balance capable of measuring forces up to ~4N (400 gram mass equivalent with a precision of 0.01gram). (PASCO model SE 8709 or Ohaus scout STX 422)
- Lab stand (PASCO model ME-9355)
- Hookup wires with banana plug connectors (PASCO model SE-9750 (red) & SE-9751 (black))
- Current Balance Accessory (PASCO model SF-8608)

The Pasco current balance assembly consists of a main unit, 6 current loop PC boards and a magnet assembly with several magnets to vary the magnetic field. Figure 3 shows various components of the current balance assembly. The current balance accessory unit consists of a rotating current balance accessory unit and a magnet assembly. Figure 4 shows components of the current balance accessory unit. Use figures 3 and 4 to identify various components in configuring your experiment.

Experiment 4: Force on a current carrying wire in a magnetic field

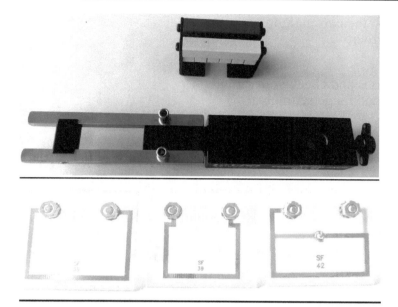

Fig.3 The Basic current balance; Main unit, Current loops and Magnet assembly (PASCO SF 8607)

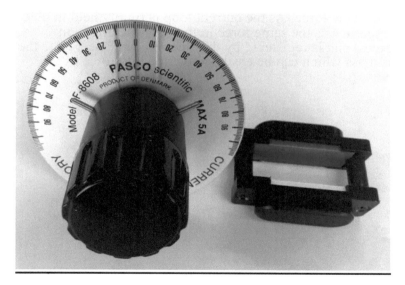

Fig.4 The Current balance accessory to measure the angular dependence of the force (PASCO SF 8608)

Equipment Safety

- Before switching the power supply, review and follow all precautions in using the power supply.
- Turn off the power supply before changing current loops.
- Do NOT keep the current in the current loops on for a long time. Quickly take measurements and move on.

Experiment 4: Force on a current carrying wire in a magnetic field

- Do NOT exceed the current rating for the current loops. If you exceed the current rating, it will cause permanent damage to the current loops.
- Have the instructor check your circuit.

Before you begin, the instructor will:
- Review electrical safety precautions and check your circuit.
- Show you meter input terminals and range settings.

The objects of this experiment are
I. Measure the magnetic force on a current carrying wire.
II. To verify that the force varies linearly with the length L, and current I, of the current the loop.
III. To see if the angular variation of the force on a current carrying wire is sinusoidal according to the expression C.
IV. Get familiar with the Right hand rule to find the direction of the magnetic force on a current carrying wire.

General principle of operation:

In this experiment, to measure the force on a current carrying wire, we use a top loading electronic balance to measure the change in mass and thereby force as a function of current in a wire-loop. The magnet assembly will be placed on the top of an electronic balance and the wire-loop is located inside the magnet assembly as shown in the figure 5. When a current is passed through the current loop, the loop experiences a force due to the magnetic field. See figures 2 and 6 for the direction of the force on the current carrying wire and the magnet assembly. Use the right hand rule to verify the direction of the force. As the current loop is rigidly held in place by the lab stand the magnet assembly experiences the same force in the opposite direction (Newton's third law, action reaction forces) and is registered by the electronic balance. Note the electronic balance registers the force as mass which can be easily converted into force by multiplying by acceleration due to gravity, g.

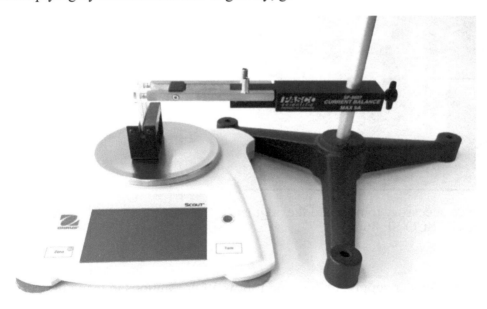

Fig 5: The basic current balance setup with current loop, magnet assembly and balance.

| Experiment 4: | Force on a current carrying wire in a magnetic field |

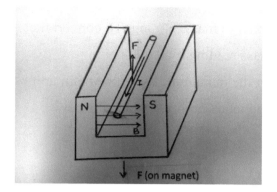

Fig 6: Force on current carrying wire placed in a magnetic filed and the reaction force on the magnet.

PROCEDURE

To measure the force on a current carrying wire as a function of current and length use the procedure below.

Setup the current balance as shown in figure 5.

1. Mount the main unit of the current balance (SF-8607) on the lab stand as shown in fig. 5.
2. Plug in the 1.2 cm current loop (SF 40) into the ends of the arm of the main unit.
3. Place the Magnet assembly on the electronic balance (Ohaus Scout STX 422).
4. Lower the current loop slowly into the magnet assembly and position the lab stand so that the horizontal portion of the conductive foil on the current loop passes through the poles region of the magnets (see figs 5 and 6).
5. Make sure that the current loop is not touching the sides or bottom of the magnet assembly and is parallel to the magnet poles. It will be easier to rotate the magnet assembly than moving the stand to get this alignment.
6. Connect the main arm of the current balance to the DC power supply (SF 9584). Have an ammeter in the circuit to measure the current. You can also use the built-in ammeter in the power supply. See sketch of the circuit given in figure 7.
7. The DC power supply (SF9584) allows adjusting the current and voltage independently using 2 separate knobs. To begin, turn the voltage control knob clock-wise all the way (to maximum) and the current control knob counter-clock-wise all the way (to minimum).

| **Experiment 4:** | **Force on a current carrying wire in a magnetic field** |

8. Note in this experiment, we are recording force in mass units as read by the electronic balance. You also realize that the force is proportional to mass, F = mg. To convert the mass in grams to force in Newtons, multiply the mass reading by 0.0098 N/g.

9. To change the current loop, loosen the set screw on the main unit; swing the main unit such that the current loop moves away from the magnet assembly. Attach the new current loop and repeat step 5 to align and take data.

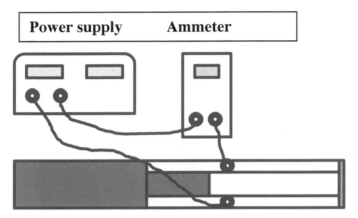

Fig. 7: Electrical connections to the current balance.

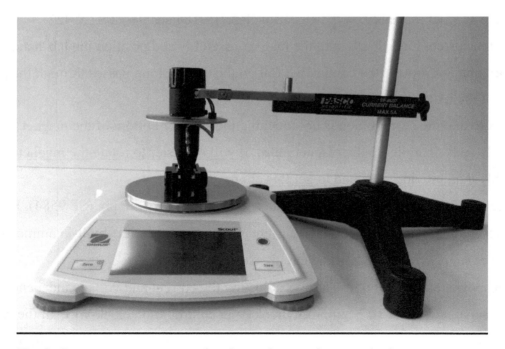

Fig 8: Setup to measure angular dependence of magnetic force on a current carrying wire.

Experiment 4: Force on a current carrying wire in a magnetic field

To measure the force on a current carrying wire as function of the angle between magnetic field and current (length) use the procedure below.

Setup the current balance along with current balance accessory as shown in the figure 8.

1. Mount the main unit of the SF-8607 current balance on a lab stand as shown in the fig. 8.
2. Plug in the current balance accessory unit SF-8608 into the main unit.
3. Place the compatible magnet assembly on the electronic balance (Ohaus scout STX 422).
4. Lower the unit slowly into the magnet assembly and position the lab stand so that the coil portion of the accessory unit is inside the pole region of the magnets and is not touching on the sides.
5. Rotate the coil and check that the accessory unit is not touching on sides of the magnet assembly or pan of the balance.
6. Connect the main arm of the current balance to the DC power supply (SF 9584). Have an ammeter in the circuit to measure the current. See the circuit diagram given in fig 7.
7. The DC power supply (SF9584) allows adjusting the current and voltage independently using 2 separate knobs. To begin, turn the voltage control knob clock-wise all the way (to maximum) and the current control knob counter-clock-wise all the way (to minimum).
8. Note in this experiment, we are recording force in mass units as read by the electronic balance. You also realize that the force is proportional to mass, $F = mg$. To convert the mass in grams to force in Newtons, multiply the mass reading by 0.0098 N/g.

| Experiment 4: Force on a current carrying wire in a magnetic field: Introduction |

Name: _____   Section/ Group: _____

Lab Partners: _____   Instructor's Initials: _____

Introduction:

Experiment 4: Force on a current carrying wire in a magnetic field: Data

MEASURING THE FORCE

1. <u>Measuring the force as a function of current in the current loop:</u>

Follow the instructions given under procedure (page 47) to set up the current balance. Attach the SF-40 (1.2 cm) current loop to the main unit and complete the wiring to connect the current loop to the power supply. Have your instructor check the setup and the electrical circuit. Turn on the power supply. At this point the ammeter should read zero current. Tare any mass reading on the electronic balance using the tare button on the balance. Turn the voltage control knob clockwise all the way (maximum) and the current control knob counter-clock-wise all the way (minimum). To take data, slowly increase the current by rotating the current knob. Record the current and corresponding mass readings on the electronic balance in table 1.

Take a total of 10 data points by increasing the current from 0 to 5.0A in steps of 0.5A. Repeat data for 3 more loops [SF-37 (2.2 cm), SF-38 (4.2 cm), SF-42 (8.4 cm)].

Table 1: Force as a function of current.

Magnetic field strength _____

Loop	SF-40 (1.2 cm)		SF-37 (2.2 cm)		SF-38 (4.2 cm)		SF-42 (8.4 cm)	
Current	Mas	Force	Mas	Force	Mas	Force	Mas	Force
A	g	N	g	N	g	N	g	N
0.5								
1.0								
1.5								
2.0								
2.5								
3.0								
3.5								
4.0								
4.5								
5.0								

Experiment 4: Force on a current carrying wire in a magnetic field: Data

2. <u>Measuring the force as a function of length, L of the current loop:</u>
   There is no need to make additional measurements to obtain this data. Use the data from table 1, rearrange the data and complete the data table-2.

Table 2: Force as a function of length of the current loop.

Loop Length cm	I = 1.0A Force N	I = 2.0A Force N	I = 3.0A Force N	I = 4.0A Force N	I = 5.0A Force N
1.2					
2.2					
4.2					
8.4					

3. <u>Measuring the force as a function of the angle θ between magnetic field and current I (length L).</u>

Follow the instructions given under procedure (page 49) to set up the current balance. Attach the angular accessory to the main unit and complete the wiring to connect the current loop to the power supply. Have your instructor check the setup and the electrical circuit. Turn on the power supply. At this point the ammeter should read zero current. Tare any mass reading on the balance using the tare button on the balance. Turn the voltage control knob clockwise all the way (maximum) and the current knob counter-clock-wise all the way (minimum). To take data, set the dial on the unit to 0° and set the current to a value of 2A in the coil and record mass reading on the electronic balance. Slowly rotate the dial clockwise in 10° increments and obtain mass reading at each new setting. Record the angle and corresponding mass readings on the electronic balance in table 3.

Take a total of 18 data points by increasing the angle in steps of 10° from -90° to 90°.

Experiment 4: Force on a current carrying wire in a magnetic field: Data

Table 3: Angular dependence of magnetic force on a current carrying wire:

Current in the loop _____   Magnetic field strength _____

Angle θ	Mass	Force = 0.0098 m	Angle θ	Mass	Force = 0.0098 m
0			0		
-90			10		
-80			20		
-70			30		
-60			40		
-50			50		
-40			60		
-30			70		
-20			80		
-10			90		

Graphs
1. Use the data in table-1 and plot force F (vertical axis) vs Current I (horizontal axis). Curve fit to a linear function and obtain the <u>slope</u> and <u>error slope</u>. Pay attention to the units.

2. Use the data in table-2 and plot force F (vertical axis) vs Length L (horizontal axis). Curve fit to a linear function and obtain the <u>slope</u> and <u>error slope</u>. Pay attention to the units.

3. Use the data in table-3 and plot force F (vertical axis) vs Angle θ (horizontal axis). Curve fit to a Sine function.

Experiment 4: Force on a current carrying wire in a magnetic field: Analysis

Analysis:

(1) Does the force vary linearly with the current in the loop? To what percent precession? (Use the error slope obtained for the plot to answer this question)

(2) Does the force vary linearly with the length of the loop? To what percent precession? (Use the error slope obtained for the plot to answer this question)

(3) Refer to equation C, can your data be explained using this equation? Explain.

(4) From the slope of **F** vs **I** plot, for each current loop, calculate the magnetic field strength **B**. Calculate **B** (average) and Δ**B** (average).
B = Slope/L. Don't forget to convert the length into meters.

(5) Does the value of B (average) agree with the magnetic field used in the experiment? Use a 10% error in the Gauss Meter measurements.

(6) Refer to equation C, can your data of F vs θ be explained using this equation? Explain.

(7) What is the maximum force possible for the SF8608 assembly? Use B= 375G and l = 0.121m. Compare it with the maximum force in the experiment.

Summary and Conclusions

Review Questions

1. A 10.0cm long wire carrying a current of 4.5A is placed in a magnetic field B = 0.3T. Calculate the force in the wire;
   a) When the magnetic field and the current in the wire point in the same direction.
   b) When the magnetic field is perpendicular to the length of the wire.
   c) When magnetic field makes an angle of 35° with the wire.

2. In an experiment similar to the one you did to determine the force on a current carrying wire, a current of 3.8A is registered in the current loop. If the length of the loop, that is perpendicular to magnetic field, is 4.5cm and the magnetic field strength is 0.098T calculate the change in mass registered on the balance.

3. An 8.5 cm long wire, carrying a current of 3.8A is placed in a magnetic field of 0.15T. If the wire experiences a force of 0.025N. Calculate the angle between magnetic field and length of the wire.

4. In an experiment similar to the one you did to determine the force on a current carrying wire, the slope of force vs current plot is 1.85×10^{-3} N/A for a 2.2cm loop. In the experiment if the magnetic field is perpendicular to the loop, determine the magnetic field strength in Tesla.

Answers: 1(a) 0N, 1(b) 0.135N, 1(c) 0.0774N.
          2. m = 1.71 mg.   3. θ = 31°   4. B = 0.0841T

Experiment 5 Focal Length of a lens and Image formation

Converging Lens (Convex Lens)

A glass lens which is thicker at its center than at its rim will converge incoming parallel light rays to an observable focused spot at the primary focal point F. The focal length f is the distance from F to the center of the lens. For a *converging* lens (Fig. 1), f is defined as positive. Since light ray paths are reversible, rays coming from the focal point F toward the lens will be made parallel after transmission through the lens.

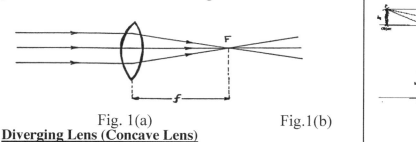

Fig. 1(a)   Fig.1(b)

Diverging Lens (Concave Lens)

A *diverging* lens (Fig. 2), which is thinner at its center than at its rim, will diverge incoming parallel rays so that they appear to have come from a focal point F to the left of the lens, which is not directly observable. The focal length f is a negative number.

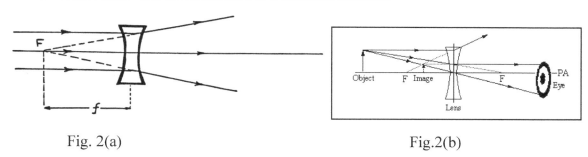

Fig. 2(a)   Fig.2(b)

A thin lens is one whose thickness is much less than its focal length. A strong lens is one that bends rays more sharply and therefore has a short focal length. The power of a lens in diopters is defined as the inverse of its focal length in meters.

$$P = \frac{1}{f} \qquad (A)$$

diopters = 1/focal length in meters

Since rays from a small, distant object are almost parallel, the focal point of a converging lens is easily found by locating the image position F of a distant object and measuring the distance f to the lens. When a thin lens is reversed, the same focal length will be observed.

Since the focal point of a diverging lens is not directly observable, its focal length f_2 must be found indirectly by putting the diverging lens in contact with a converging lens whose focal length $f_1 < |f_2|$. In combination they will have a positive focal length f_{12} which can be directly measured. The powers of the two lenses in contact add: $P_{12} = P_1 + P_2$.

Experiment 5   Focal Length of a lens and Image formation

Therefore,
$$\frac{1}{f_{12}} = \frac{1}{f_1} + \frac{1}{f_2} \tag{B}$$

OR

$$\frac{1}{f_2} = \frac{1}{f_{12}} - \frac{1}{f_1}. \tag{C}$$

f_1 – focal length of converging lens
f_2 – focal length of diverging lens
f_{12} – focal length of combination

f_2 is easily calculated from measured values of f_1 and f_{12}. f_2 will be a negative number.

Image Formation

The rays from real objects at distances d_o will be converged by a lens. Therefore they will be focused at distances d_i to form an image. Fig. 3 shows two ray diagram for a real object at a distance d_o from the lens brought to a focus at a real image at a distance d_i from the lens.

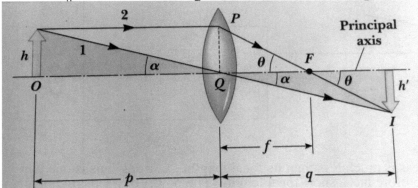

Fig. 3

The relation between f, d_o (p) and d_i (q) for both converging and diverging thin lenses is:

$$\frac{1}{f} = \frac{1}{d_o} + \frac{1}{d_i} \quad \text{or} \quad f = \frac{d_i \cdot d_o}{d_i + d_o} \quad \text{(thin lens equation)} \tag{D}$$

Note that if d_o is very large, $\dfrac{1}{f} = \dfrac{1}{\infty} + \dfrac{1}{d_i} = 0 + \dfrac{1}{d_i}$ (E)

and so $d_i = f$, as it must be by definition.

If we denote the distance between the object and the image as L, then

$$L = d_o + d_i \tag{F}$$

If we take the height of the object as h_o and that of the image as h_i (see Fig. 3), the lateral magnification m is defined as

$$m = -\frac{h_i}{h_o} \quad (G)$$

The predicted lateral magnification is

$$m = -\frac{d_i}{d_o} \quad (H)$$

Since real images have positive distances $+d_i$, the minus sign tells us that a real image should be inverted, as it is in Fig. 3.

The OBJECTIVES of the experiment are:
I. Learn to distinguish quickly between converging and diverging lenses by their shape.
II. Measure the focal lengths of converging and diverging lenses by forming images of a distant object.
III. Measure the focal length and magnification of a converging lens by measuring the object and image distances.
IV. Understand the relation between the focal length, f object distance, d_o image distance, d_i.

EQUIPMENT: 5 - thin lenses lettered A, B, C, D, E,G; Optical bench and one f = 10 cm lens.

1. Sorting Lenses by Shape
Place a lens tissue over the lens and feel the thickness of each lens between your thumb and forefinger.
List lenses by letters and indicate if the lens is converging or diverging.

2. Measuring the Focal Length Using a Distant Object
Position yourself far away (~ 20 meters) from the light source in the room.
Mount a small cardboard screen on the meter stick. You can also use wall as the screen and butt the meter stick to the wall.
Point the stick at the distant light source, hold the lens on the stick between the screen and the light source, and record the lens and screen positions when you get a focused image of the source on the screen.
(a) Record the positions of the screen and lens and determine the focal length f of all the <u>converging</u> lenses. Estimate the reading error, Δf in each case and record
(b) Pick the converging lens with smallest focal length (highest power), use the black lens holder and put each of your <u>diverging</u> lenses in contact with a converging lens. As described above, in step(a), obtained the focal length of the lens combination, f_{12}.
RECORD the data. CALCULATE the focal lengths f_2 of the diverging lenses from f_1 and f_{12}.

3. Measuring the focal length from object and image positions, Real Image Positions and Magnifications

Set up the optical bench as in Fig. 4. Place the light source (object) on the left end of the bench, the white screen on the right end, and the lens somewhere in between. Move the lens and get a sharp image of the object on the screen. Measure do = the distance between object and the lens and di = distance between the lens and the image. Calculate the focal length. Compare the values with the results obtained in steps 2 and 3.

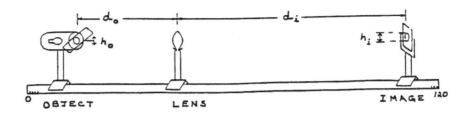

Fig.4

You will use PASCO optic bench and accessories. The lens to use here is a converging lens whose focal length is approximately 10 cm. Set up the optical bench as in Fig. 6 with the readings increasing as you go from left to right. Place the light source (object) on the left end of the bench, the white screen on the right end, and the lens somewhere in between. Move the lens and get a sharp image of the object on the screen. If the distance between the light source and screen (i.e., object-image distance, L) is sufficiently large, there will be two lens positions that give sharp images. These two lens positions are *conjugate* focal positions, i.e., the set of object and the image distances are interchanged for these two positions. One will give a magnified image; the other will give a demagnified image. If L is too small, however, the lens will not be able to converge the rays and a sharp image will not be found. Familiarize yourself with the optic bench and components (given in Fig. 4) before proceeding further.

- Find out the largest and smallest values of L that can give you two focused images (small and large) on the screen.

Start with the largest possible value of L (approximately the length of the optical bench) and RECORD the object and image positions (P_o and P_i) as read on the optic bench. It is convenient to keep $P_o = 0$ cm and $P_i = 100$ cm or 110 cm (a round number), so that it is easy to compute d_o and d_i.

- RECORD the two lens positions (P_L on the bench) which give sharp images, the size of the object and image (h_o and h_i) in each case, and also the nature of the image (large or small; inverted or erect). Calculate d_o, d_i, d_i/d_o, h_i/h_o, and f for these two lens positions.

- Check to see if the calculated f is close to 10 cm and the magnification m is as expected from equations (G) and (H). Continue with the experiment if you have agreement. Record also the reading errors, Δd_o and Δd_i for these two measurements and give your basis for their estimate.

- PLOT this data point on a linear graph with d_o along x-axis and d_i along y-axis.

- **Choose the <u>smallest</u> value of L**: RECORD P_o, P_i, P_L, h_o, and h_i for which two sharp images can be found. Calculate the other quantities on the table and plot this data point also on the same graph.

- CHOOSE four or five additional values of L and RECORD P_o, P_i, P_L, h_o, and h_i for these values of L. Compute d_o and d_i and plot each of these data points on the same graph <u>as you take the data</u>. Record additional data, if needed to define the nature of the plot completely). Note down in each case whether the image is erect or inverted.

Graphs

(a) Create a table in Logger Pro with 8 columns as in Table 5 on page 67 and enter data from the Tables 3, 4 and 5.

| d_o | d_i | L | h_i | h_o | d_i/d_o | h_i/h_o | f |

(a) Plot d_i (vertical axis) versus d_o (horizontal axis) and fit the data with the automatic curve fitting option using the formula:

$$g(x) = (x * A) / (x - A).$$

Note: g(x) is the y-value; in this case it is d_i. From the thin lens equation (D)

$$\frac{1}{d_i} = \frac{1}{f} - \frac{1}{d_o} \quad \text{or} \quad d_i = \frac{d_o * f}{d_o - f}.$$

$(g(x) = d_i = d_o * f / (d_o - f)$ with $x = d_o$ and $A = f$.)

This is the equation you are trying to fit to the observed data. Comparing the equation for g(x) and d_i, you can notice that the parameter A should be the focal length f.

(b) Plot L (vertical axis) versus d_o (horizontal axis) and do a curve fitting to the data using the formula:

$$g(x) = x + (A * x) / (x - A). \quad g(x) = L = d_o + d_i = d_o + d_o f/(d_o - f); A = f; x = d_o$$

Do you understand this formula?

(c) Plot $|h_i/h_o|$ (vertical axis) versus d_i/d_o (horizontal axis) and do a curve fit and select linear fit.
   Find the slope and the error in the slope for graph (c).

Experiment 5. Focal Length of a Lens   Introduction

Name: Section/ Group:

Lab Partners: Instructor's initials:

Experiment 5: Focal Length of a Lens   Data

1. Sorting Lenses by Shape

Converging: Diverging:

2. Measuring the Focal Length Using a Distant Object

Approximate distance between screen and light bulb: _____

Note Δf is the reading error in determining the focal length.

Table-1

Lens	Screen Position	Lens Position	f	Δf

Lens combinations:

Table-2

Lens Combination	Screen Position	Lens Position	f_1	f_{12}	f_2

Show calculations for f_2:

Experiment 5: Image Formation

Table 3: Real Image Positions and Magnifications (measured with the optic bench set-up)

Large value of L = _____ actual f value = _____ Δd_o = _____ Δd_i = _____

Δd_o and Δd_i are the reading errors in d_o and d_i respectively.

Nature of Image	P_o	P_L	P_i	h_i	h_o	d_i	d_o	d_i/d_o	h_i/h_o	f
large, inv										
small, inv										

(Show focal length, f calculations)

Table 4: Small value of L = _____

Nature of Image	P_o	P_L	P_i	h_i	h_o	d_i	d_o	d_i/d_o	h_i/h_o	f
large, inv										
small, inv										

(Show calculations for f)

Table 5:

Nature of Image	L	P_o	P_i	$P_{L(lens)}$	h_i	h_o	d_i	d_o	d_i/d_o	h_i/h_o	f
large, inv											
small, inv											
large, inv											
small, inv											
large, inv											
small, inv											
large, inv											
small, inv											
large, inv											
small, inv											

| **Experiment 5.** | **Image Formation** | **Analysis** |

Analysis

Refer to the data taken with the distant object.

1. Refer to focal lengths provided by the instructor/s for the respective lenses:
   Are the f-values the same for a given lens? If not, what is the deviation? What could be the cause of the deviation?

2. You have a concave lens whose focal length has an absolute value of 10 cm and several convex lenses with f = 5 cm, 8 cm, 12 cm, 15 cm, and 20 cm. Which of the convex lenses can be used to determine the focal length of the concave lens? Give your reason.

Refer to the data taken with the optical bench set-up.

3. Do all your values of d_o and d_i yield comparable values of focal length, f? Within the experimental error? Explain with reference to your data.

4. What is the best value of f (av)± Δf (av) obtained by averaging all the data from tables 3, 4, and 5? Δf (av) is the overall error in f for your set of data.
   <u>Show calculations f (av) and Δf (av).</u>

| Experiment 5. | Image Formation | Analysis |

5. What is the expected value of f for the lens you used? [See the label on the lens.] Does your value of f agree with the expected value or not? Explain. What is the percent difference between the expected and the experimental values?

6. Refer to the graphs and answer the following questions.

   (a) Refer to your graph of d_i versus d_o. Does increasing the object distance always decrease the image distance? Support your answer with numbers from your data. (Tables 3-5)

   (b) At what value of d_o are d_i and d_o equal? Compare this value of d_o with f. (Mark this point on your plot)

7. Refer to the plot of L versus d_o. At what value of d_o is L minimum? Compare this value of d_o with the answer to 6 (b) and the focal length. (Mark this point on your plot)

8. Does magnification m depend linearly on d_i/d_o? At what value of d_o is the |magnification| = 1? Compare this with the $d_o = d_i$ value from 6(b).

(5) Are the **real** images of a converging lens always inverted? Circle YES or NO.

Summary and Conclusions

| **Experiment 5:** | **Review Questions** |

(1) Answer CONVERGING, OR DIVERGING.
    (a) a lens thicker in the middle than at edges
    (b) a lens which magnifies close objects.
    (c) a lens with $f = +10$ cm put in contact with a lens of $f = -5$ cm.

(2) Given $f = +12$ cm, calculate d_i, L, and m when d_o is (a) 36 cm; (b) 24 cm; (c) 18 cm. (Note: (a) and (c) are conjugate focal positions.)

(3) Is d_o larger or smaller than d_i when you have a real, magnified image?

(4) A converging (lens $f = +12$ cm) in contact with a diverging lens gives a combined focal length of $+36$ cm. Calculate the focal length of the diverging lens.

Answers:
    (1) (a) converging; (b) converging; (c) diverging
    (2) (a) $d_i = 18$ cm, L = 54 cm, m = -0.5; (b) $d_i = 24$ cm, L = 48 cm, m = -1;
        (c) $d_i = 36$ cm, L = 54 cm, m = -2.0
    (3) smaller.   (4) -18 cm.

Experiment 6: Polarization of light and verification of Malus' law

Polarization of light

Visible light is a small part of the electromagnetic spectrum consisting of low energy Radio waves at one end of the spectrum to high energy Gama rays at the other end of the spectrum. Electromagnetic waves are waves with oscillating electric and magnetic fields, oscillating in mutually perpendicular planes. EM waves are transverse waves with EM field oscillations perpendicular to the direction of wave propagation. These waves can travel through solids, liquids gases and vacuum. All EM waves travel through vacuum with a constant velocity 3×10^8 m/s, known as velocity of light, c a constant. Apart from exhibiting normal wave properties such as reflection, refraction and interference, transverse waves such as EM waves also exhibit Polarization.

Polarization:

To understand the phenomena of polarization consider the light emitted by an incandescent light bulb. In an incandescent light bulb, the rapidly moving electrons (charges) in the hot filament give rise to electric and magnetic fields and thus generating an EM wave. The EM wave generated by a single electron will have electric and magnetic fields confining to two mutually perpendicular planes and their orientation remains constant with time as the wave propagates. However, the emission in an incandescent bulb is due to large number of electrons and therefore it can be thought of having several mutually perpendicular planes with electric and magnetic fields and the light is said to be unpolarized. Figure 1 shows a representation of an unpolarized light with E-field orientation in several random planes. It is understood that B-field also behaves in the same manner.

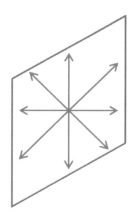

Fig.1: Representation of an unpolarized light beam.

When an unpolarized light passes through a polarizer it produces a light with E-field oscillations in a single plane that is parallel to the axis of the polarizer and the light said to be

Experiment 6: Polarization of light and verification of Malus' law

plane polarized. A polarizer eliminates all other E-field vibrations that are not parallel to the axis of the polarizer. When an unpolarized light is incident on an ideal polarizer the intensity of the transmitted plane polarized beam would be exactly 50%. This is because all E-filed vibrations in an unpolrized light can be resolved in to two components; one parallel to the axis of the polarizer and the other perpendicular to the polarizer axis, and only parallel component (i.e 50%) get transmitted.

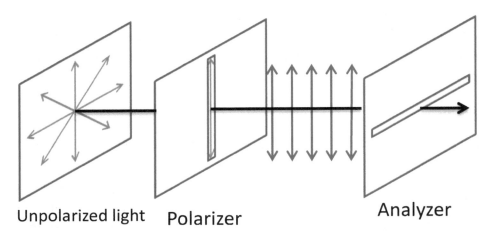

Fig. 2: When an unpolarized light is incident on a polarizer the transmitted light is plane polarized.

Malus' law:

When a plane polarized light is incident on a second polarizer (normally called analyzer), the intensity of the transmitted light depends on the angle ϕ between the axis of the polarizer (same as plane of polarization) and the axis of the analyzer. If the plane of polarization is parallel to the axis of the analyzer ($\phi = 0°$) the light will be transmitted un-attenuated and if the analyzer axis is perpendicular to the plane of polarization ($\phi = 90°$) the incident light will be completely attenuated with no transmission. This is shown in figure 2. The transmitted electric field, **E** can be expressed in terms of the incident electric filed **E₀** as

$$\mathbf{E} = \mathbf{E_0} \cos \phi \tag{A}$$

Since the intensity is square of the electric filed

$$I = I_0 \cos^2 \phi \tag{B}$$

The expression B gives a relation between the incident and transmitted intensities of a polarized light and is known as Malus' law.

Objectives of the experiment:

Experiment 6: Polarization of light and verification of Malus' law

1. Learn to configure a measurement using PASCO's Capstone software.
2. To see if the laser light is polarized.
3. To verify Malus law of polarization.

Equipment:

1. Basic Optic bench (60 cm) OS-8541
2. Red diode laser S-8525A
3. Polarization analyzer OS-8533A (2)
4. High Sensitivity light sensor PS-2176
5. PASCO Capstone software (already installed on the computer)

Laser Safety:

Laser is an acronym for Light Amplification by Stimulated Emission of Radiation. This light has several special and desirable properties such as high brightness, single wavelength, high degree of collimation, and coherence. Because of its brightness and small spot size, you should NEVER LOOK DIRECTLY INTO THE LASER! But it is safe to see the beam projected on a non-reflective screen (as in laser shows). A diode laser or semiconductor laser contains a p-n junction. Every compact disc (CD) player has a semiconductor laser in it. Fiber optic communication uses the light from semiconductor lasers. The diode laser that you will use consists of an AlGaInP quantum well structure. Its output wavelength ranges from 650 to 680 nm and the output power is < 5 mW (class IIIa). The location of the beam can be adjusted by turning the Horizontal and vertical adjustment screws on the back of the Laser.

Procedure:

Configuring PSCO's CAPSTONE software to take data:

In this experiment we use the high sensitivity light sensor interfaced to a computer to measure the light intensity as it passes through a polarizer and analyzer. The PASCO's CAPSTONE software is a new generation software, similar to the PASCO's Data Studio (we will be using Data Studio in experiments 7 and 8), for the computer interface and data acquisition purpose. PASCO's Pasport sensors work only with the CAPSTONE software. In the current experiment we will connect the high intensity light sensor directly to the computer's USB port using PASCO Airlink. There is no need to use an interface. Configure the Capstone software using following the steps.

1. Launch the Capstone software by double clicking the Capstone icon on the computer.
2. The program will open with a screen as shown in figure 3(a)

Experiment 6: Polarization of light and verification of Malus' law

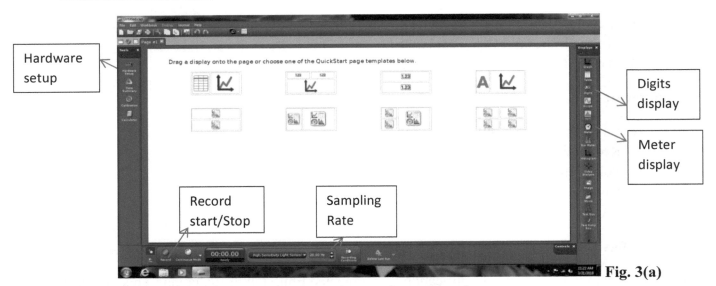

Fig. 3(a)

3. Get familiar with various icons and controls on the software.
4. Click on the Hardware tab to see if the light sensor is connected. It should open a window showing the light sensor as shown in the figure. (Fig. 3(b))

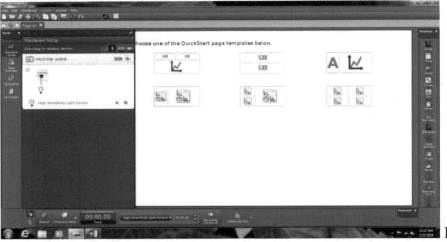

Fig. 3(b)

5. On the right, click on the Meter icon to create a meter display to read the light intensity.
6. On the right, click on the Digits icon to create a Digital display to read the light intensity. This makes convenient read light intensity in digital format. See the figure 3(c).

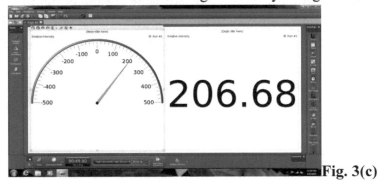

Fig. 3(c)

Experiment 6: Polarization of light and verification of Malus' law

7. Click <on Select measurement> on top left corner of the meter (also on the digital display) and choose relative intensity.
8. Set the Sampling rate to 500Hz.
9. Now, you can click Record button to start recording (measuring) the light intensity.
10. This completes setting up of the Capstone to take data. At this point if light is shined on the light sensor the meter as well as the Digital display should react and indicate a value.

Setting up the Optical bench:

Setup the optical bench as shown in figure 4. Note that components are separated for clarity. At the time of the experiment they need to be pushed together to minimize intensity loss. Get 1

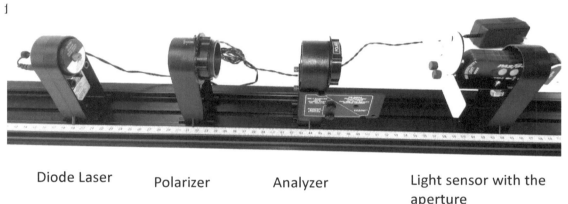

Diode Laser   Polarizer   Analyzer   Light sensor with the aperture

Fig 4: The Optical bench setup for the polarization measurements.

a) The red diode laser produces a laser beam in 650-680 nm range. **Figure 5**

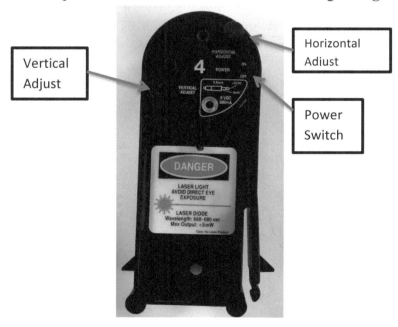

Experiment 6: Polarization of light and verification of Malus' law

b) Polarizer has an axis and it will allow the light with E-filed vibrations parallel to the axis to pass through while eliminating all other vibrations. By rotating the polarizer, one can rotate the axis of polarization.

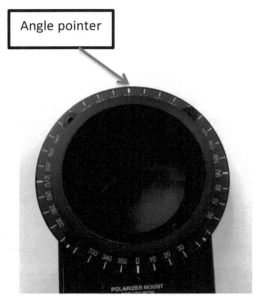

Fig. 6: Polarizer/Analyzer.

c) Analyzer is identical to the polarizer in all aspects. (Note in the figure 4 the analyzer is shown along with a Rotary motion sensor. However, we will not be using the rotary motion sensor in our experiment.
d) Light sensor detects the amount of light incident on it. In front of the light sensor there is an aperture bracket. One can control the amount of light to the light sensor by choosing different slits on the aperture brackets.

Experiment 6: Polarization of light and verification of Malus' law

Name: Section/Group:

Partner's Name: Instructor's initials:

Introduction:

Experiment 6: Polarization of light and verification of Malus' law

To check if the laser light is polarized:

To check if the laser light is polarized we will use the diode laser, analyzer and a light sensor. If the laser light is polarized, we should see a variation in the light intensity, as detected by the light sensor,. This is because a polarized laser light will have E-field vibrations in only one plane. The light sensor would detect maximum intensity if the axis of the analyzer is parallel to this plane and minimum intensity when it is perpendicular.

Procedure: (**This experiment can be done with the room lights ON**)

1. Set up the optic bench as described in page 77.
2. Remove the analyzer and move all the components close to each other to minimize the intensity loss. (Since polarizer and analyzer are identical we use the polarizer as analyzer)
3. Power the diode laser. Use horizontal and vertical adjustments to on the laser to make the beam incident on the vertical aperture attached to the light sensor.
4. Configure the PASCO's Capstone software to read the light intensity as described in pages 75 and 76.
5. Rotate the polarizer to 0° and record the light intensity.
6. Slowly rotate the polarizer from 0° to 180°, insteps of 15° and record the light intensity.
7. Record data in table 1 below.
8. Plot light intensity (vertical axis) vs the angle (horizontal axis) using LoggerPro.

Table 1:

Angle (degrees)	Light intensity (arb. units)	Angle (degrees)	Light intensity (arb. units)

1. Plot light intensity (vertical axis) vs the angle (horizontal axis) using LoggerPro (plot 1).

Experiment 6: Polarization of light and verification of Malus' law

Verification of Malus' law of polarization:

When a plane polarized light is incident on an analyzer, the intensity of the transmitted light depends on the angle ϕ between the axis of the polarizer (also the plane of polarization) and the axis of the analyzer. In this experiment we use a polarizer to polarize the laser light beam and measure the intensity of the light detected by the light sensor as a function of angle ϕ between analyzer axis and the axis of the polarizer (the plane of polarization). As the laser beam is naturally polarized, the role of polarizer is to force the plane of polarization along polarizer axis.

Procedure: (This experiment can be done with the room lights ON)

1. Set up the optic bench as described in page 77
2. Move all the components close to minimize the intensity loss.
3. Power the diode laser. Use horizontal and vertical adjustments on the laser to make the beam incident on the vertical aperture attached to the light sensor.
4. Configure the PASCO's Capstone software to read the light intensity as described in pages 75 and 76.
5. Place the polarizer at an angle where the light intensity was maximum in the previous experiment. If needed, rotate the polarizer by ± few degrees to locate the maximum intensity position (angle) and keep it there. This step is to align polarizer axis parallel to plane of polarization of the diode laser beam.
6. Choose appropriate aperture so that the relative light intensity readings on the light sensor (in Capstone software) are high.
7. Rotate the analyzer to Zero degrees and record the light intensity in the preliminary data.
8. Rotate the analyzer by $90°$ degrees and record the light intensity in the preliminary data.
9. Slowly rotate the analyzer from $0°$ to $360°$, insteps of $15°$ and record the light intensity.
10. Record data in table 2 below.
11. Plot light intensity (vertical axis) vs the angle (horizontal axis) using LoggerPro.

Preliminary observations:

Measures the relative light intensity of the laser beam with

(i) no polarizer or analyzer in the optical path,
(ii) the polarizer only in the optical path and
(iii) with both polarizer and analyzer in the optical path.

This is to observe the decrease in intensity when the polarizer and analyzer are introduced in to the optical path.

Experiment 6: Polarization of light and verification of Malus' law

Relative light intensity of the laser beam (No polarizer or analyzer in the optical path) _____

Relative light intensity (Only the polarizer in the optical path, angle at 0°) _____

Relative light intensity (Both polarizer and analyzer in the optical path, angle at 0°) _____

Relative light intensity ($\phi = 0°$) _____   Relative light intensity ($\phi = 90°$) _____

Table 2:

Angle, ϕ (degrees)	Light intensity (arb. units)	Angle, ϕ (degrees)	Light intensity (arb. units)

1. Use the LoggerPro to plot light intensity (vertical axis) vs the ϕ angle (horizontal axis).
2. Curve fit to a Cosine function (plot 2).
3. Calculate $\cos\phi$ and $\cos^2\phi$ using calculator (Don't use the LoggerPro to calculate).
4. Plot intensity vs $\cos\phi$ (plot 3) and intensity vs $\cos^2\phi$ (plot 4).

Experiment 6: Polarization of light and verification of Malus' law

Analysis

(1) Refer to light intensity vs angle plot for the table 1, is the diode laser beam polarized? Explain.

(2) What is the shape of the light intensity vs angle plot?

(3) Refer to the data from table-2, does your data support Malus law of polarization. Explain qualitatively and then support your answer with numbers.

(4) Use the preliminary observations to answer these questions;

Compared to no polarizer or analyzer in the optical path, by what percent does the light intensity decrease when

(a) The polarizer is introduced into the optical path?
(b) The both polarizer and analyzer are introduced into the optical path?

Experiment 6: Polarization of light and verification of Malus' law

(5) Using the relative light intensity, when both polarizer and analyzer are in the optical path and at 0°, as the reference intensity use equation B to calculate theoretical intensity at $\phi = 30°$ and $\phi = 45°$. Compare the calculated values with your measurements at these angles. Do calculated and measured values agree? What is the percent difference?

Summary and Conclusions:

Experiment 6: Polarization of light and verification of Malus' law

Review Question:

1. Explain polarization of light in your own words.

2. In a polarization of light experiment an incandescent light source is used. The ratio polarized to unpolarized light intensity is
   (a) 25% (b) 50% (c) 75% (d) 100%

3. In a polarization of light experiment the relative light intensity when $\phi = 0°$ is measured to be 385. What is the expected relative light intensity when $\phi = 30°$ and $\phi = 65°$.

Answers:

2. (b) 3. 288.75 and 68.76.

Experiment 6: Polarization of light and verification of Malus' law

Experiments 7 and 8: Diffraction and Interference of Light

The OBJECTIVES of these experiments are:
    (1) to determine the wavelength of a diode laser;
    (2) to study the diffraction and interference of light waves using single slits and
    (3) to study the diffraction and interference of light waves using double slits.

Measurement of Wavelength

A diffraction grating is a plate on which very closely spaced parallel lines are scribed. The grating you will use is a plastic film (mounted on a cardboard) which is a replica of a very precisely ruled metal grating. The separation, d, between the lines must be larger than the wavelength of light, but small enough so that several of them are illuminated by the light from the diode laser. If there are N lines in a given length,

$$d = 1/N.$$

When light of wavelength λ, falls on the grating, the waves that pass through different slits interfere constructively only if the following relation is satisfied.

$$d \cdot \sin\theta_m = m\lambda \qquad (A)$$

where m is an integer called the order of diffraction, i.e., m = 1 for the first order, m = 2 for the second order, and so on; θ_m is the angle from the un-diffracted direct beam (at $\theta = 0$) to the bright spot of order m. The diffracted pattern is symmetrical; so there will be two spots corresponding to every order situated on either side of the direct beam (Fig. 1).

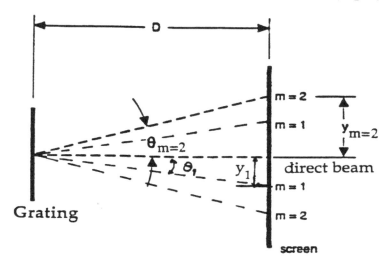

Fig. 1

If D is the distance between the grating and the screen, and y_m the distance between the direct beam and the diffraction spot of order m (Fig. 1), then $y_m/D = \tan\theta_m$ so that

$$\theta_m = \tan^{-1}(y_m/D) \qquad (B)$$

The <u>wavelength</u>, λ, of light can be computed from the known values of θ_m and d using equation (A).

Experiments 7 and 8: Diffraction and Interference of Light

Single Slit Diffraction

If parallel monochromatic light passes through a single slit whose width is comparable to or wider than the wavelength, then portions of the wavefront passing through different portions of the slit have different phases and interference can take place. This process is known as diffraction. As a result, at distances far away from the slit, one observes an intensity distribution that is peaked in the straight-ahead position and having a series of decreasing peaks symmetrically on either side of the center (Fig. 2).

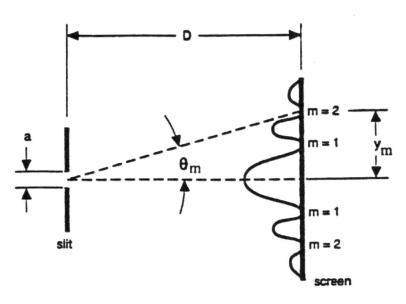

Fig. 2

The zeros between these peaks (positions of minima in the diffraction pattern) occur when

$$a \cdot \sin \theta_m = m\lambda \qquad (C)$$

where a is the slit width, θ_m is the angle from the center of the pattern to the order m minimum, λ is the wavelength of light, and m is the order (1 for the first minimum, 2 for the second minimum, ...counting from the center outward) (Fig. 2).

If the screen is sufficiently far away from the slit i.e., if D >> a, so that the angles to the various minima are small, we can approximate $\sin \theta_m = \tan \theta_m = y_m /D$. Equation (C) can then be solved for the slit width, i.e.,

$$a = \frac{m\lambda D}{y_m} \quad (m=1,2,3\ldots) \qquad (D)$$

Experiments 7 and 8: Diffraction and Interference of Light

Double Slit Diffraction

Light passing through two closely spaced narrow slits creates an interference pattern because the part of the wavefront reaching a point in the viewing screen after passing through the "top" slit differs in phase from that part passing through the "lower" slit. The resulting pattern displays a series of interference peaks, separated by intensity zeros, which are symmetrically located on either side of the central beam. The angle to the maxima (bright fringes) in the interference pattern is given by

$$d \cdot \sin \theta_m = m\lambda \quad (m = 0, 1, 2, \ldots) \qquad (E)$$

where d is the slit separation, θ_m is the angle from the center of the pattern to the m order maximum, λ is the wavelength of light, and m is the order. m = 0 for the central maximum, m = 1 for the first side maximum, m = 2 for the second side maximum, ... counting from the center out (Fig. 3.1). The intensity minima between these maxima occur at angles θ_m such that

$$d \cdot \sin \theta_m = (m+1/2)\lambda \quad (m = 0, 1, 2, 3, \ldots)$$

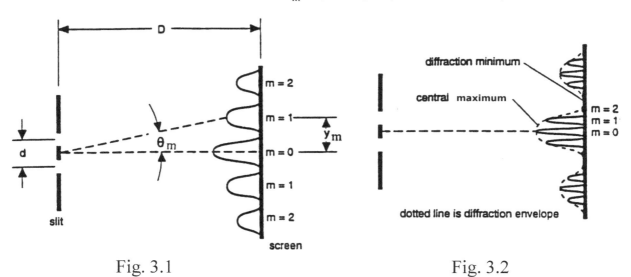

Fig. 3.1   Fig. 3.2

Since the angles are usually small, as before, we can approximate $\sin \theta_m = \tan \theta_m = y_m/D$. Equation (E) can then be solved for the <u>slit separation</u>:

$$d = \frac{m\lambda D}{y_m} \quad (m = 1, 2, 3, \ldots) \qquad (F)$$

While the interference fringes are created by the interference of the light coming from the two slits, there is also a diffraction effect occurring at each slit due to Single Slit Diffraction. This causes the envelope as shown in Fig. 3.2.

In this experiment, you will use a diode laser, diffraction grating, and different slit assemblies on an optic bench as shown below (Fig. 4). By measuring a few distances on the diffraction (and interference) pattern and using the predictions of the theory outlined above, you will be able to determine distances of the order of a few hundreds to a few tenths of a micrometer to within ~ 5% <u>or better</u> accuracy!

Experiments 7 and 8: Diffraction and Interference of Light

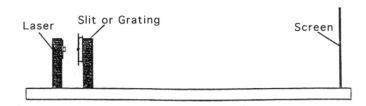

Fig. 4

Laser is an acronym for Light Amplification by Stimulated Emission of Radiation. This light has several special and desirable properties such as high brightness, single wavelength, high degree of collimation, and coherence. Because of its brightness and small spot size, you should NEVER LOOK DIRECTLY INTO THE LASER! But it is safe to see the beam projected on a non-reflective screen (as in laser shows). A diode laser or semiconductor laser contains a p-n junction. Every compact disc (CD) player has a semiconductor laser in it. Fiber optic communication uses the light from semiconductor lasers. The diode laser that you will use consists of an AlGaInP quantum well structure. Its output wavelength ranges from 660 to 680 nm and the output power is < 5 mW (class IIIa). The location of the beam can be adjusted by turning the Horizontal and vertical adjustment screws on the back of the Laser (Fig. 5).

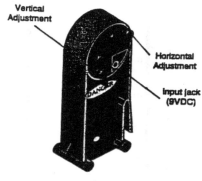

Fig. 5

EQUIPMENT
1 - 1.25 m optic bench (OS-8515)
1 - Diode Laser (OS-8525)
1 - a set of 4 diffraction gratings (in a plastic box) (SE-9361)
1 - single slit accessory on its mount (OS-8523)
1 - multiple slit accessory on its mount (OS-8523)
1 – light sensor
1 – aperture bracket for light sensor
1 - rotary motion sensor and track
1 – PASCO signal interface
1 - mount for the grating (PASCO mount modified to hold the gratings)
1 - 12" x 8" white screen (made in our machine shop) to mount on the optic bench
1 - a small white cardboard piece (or an index card) to trace the laser beam path
1 - flash lamp
1 - (for the room) scotch tape roll in a dispenser

Experiments 7 and 8: Procedure

BEFORE YOU BEGIN, the instructor will go over LASER Safety and
(a) show you how to position the laser beam parallel to the optic bench and at the center of the screen;
(b) show you how to remove and replace the slit assembly on the holder;
(c) show you the diffraction pattern from a grating, a single slit and the interference pattern from a double slit;
(d) show you how to use rotary motion and light sensor assembly to record intensity profiles using PASCO signal interface and computer.

Measurement of Wavelength

Each student should mark the pattern on their data sheet and take their own data for this part.

(1) Cover the screen with the data sheet and attach the sheet to the screen. Keeping only the laser and the screen on the optic bench (distance between them should be more than 90 cm), adjust the height and the lateral position of the laser beam so that the laser beam travels parallel to the optic bench and is centered on the screen.
(2) Place the grating with 80 lines/mm on its mount as close to the laser as possible (See Fig. 4). Line up the pattern with the line on the data sheet. Move the screen to a position where you can see two orders of diffraction on both sides of the center. RECORD the positions of the grating (note that the grating is offset from the centerline of the holder) and screen and determine D, the distance between the grating and the screen for this setup.
(3) Adjust the grating orientation so that the diffraction pattern observed on the screen is horizontal and not tilted. Locate and mark the centers (with a " + ") of the direct beam and the diffracted beams corresponding to m = 1 and m = 2 on either side of the direct beam. Label the orders below the marks.
(4) Now replace the grating with the one that has 100 lines/mm. Position your paper so that the current diffraction pattern falls on a different line on the data sheet. Mark the diffraction pattern of this grating as you did in step (3). Label the diffraction patterns so that you (and other readers of your report) can identify which pattern is from which grating.
(5) Move the grating to nearly half the D value used in step (1) and mark the spots for one of the gratings (80 or 100 line/mm) on back of the sheet.
(6) From the marks on your sheet of paper, MEASURE the distance between the two first order spots ($2 \cdot y_1$) and the distance between the two second order spots ($2 \cdot y_2$). RECORD the values of:

$$N, \quad d = \frac{1}{N}, \quad m, \quad y_m, \quad \theta_m = \tan^{-1}(y_m/D), \quad \lambda = \frac{d \cdot \sin \theta_m}{m}$$

Keep at least three significant places for d and θ, and at least four for λ.

(7) CALCULATE the average wavelength λ and compare with the average wavelength of emission from the diode laser, which is 670 nm (1nm = 10^{-9}m). Give a % difference. If the percent difference is more than 10%, check your procedures and calculations. If needed, repeat your measurement. (It is possible to get an accuracy of 1 %, if you are careful.)

REMOVE the Grating mount from the optic bench and place the gratings back in the box.

Experiments 7 and 8:   Procedure
Setup Procedure for Data Studio

The setup and procedure for studying single slit diffraction and double slit interference are identical. You will use the same setup shown in Fig. 4 with the screen replaced by a rotary motion sensor and a light sensor. Using this setup, with a PASCO signal interface and a computer, you will record the intensity profile of the diffraction of the interference pattern from the slit setup used. Before recording any data for analysis, you should experiment with the setup to familiarize yourselves. Once you become comfortable with the procedure, you will <u>plot 2 single slit and 3 double slit intensity profiles. You need to choose 1 single and 2 double slit pattern for detailed data analysis and calculations, by measuring the distances in the intensity profiles</u>. The following setup will guide you through setting up the equipment.

1. Place the single slit assembly on the optical bench about 2-3 cm from the laser (See Fig. 4). The laser should be set to the 0 cm mark on the optical bench track. For now leave the screen board on the track in front of the light and rotary motion sensors. Record the distance from the grating plate to the light sensor slit, this is your D value.

2. The wiring of the PASCO interface has already been completed for you. However, just in case the yellow plug from the rotary motion sensor should be connected to the channel 1 port and the black plug to the digital channel 2 of the PASCO interface. The light sensor plug is connected to the A port. The power switch for the PASCO interface is located on the back, and when powered on a green light in the front will indicate this setting. The PASCO-Computer connection in back is also wired prior to your experiments.

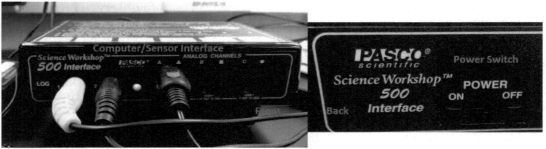

3. On the computer locate the following icon for DataStudio, and double click it.

4. Once open you will be prompted with the following GUI.

Select "Create Experiment" to setup the sensors for your experiment. Should DataStudio indicate at any time that it is not connected please alert your instructor to come and connect the proper interface.

5. After selecting "Create Experiment the following GUI will appear.

We must now setup the sensors that are used in the experiment, "Rotary Motion Sensor" and "Light Sensor".

6. Click on the Channel 1 port to get the following menu. Select "Rotary Motion Sensor", and hit OK to connect the sensor. You can also click Add sensor or Instrument and choose Rotary motion sensor from the digital sensors.

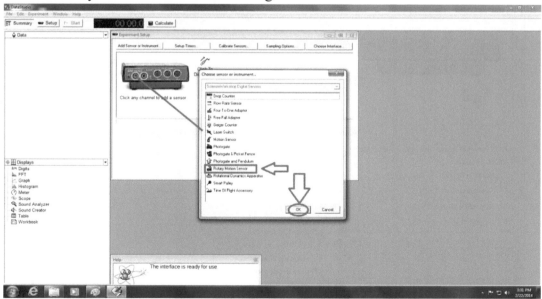

7. Double click on the sensor icon, after which the following GUI will appear.

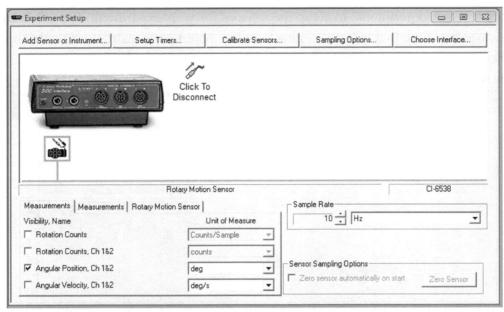

Set the sampling rate to 500 Hz and Rotary Motion Sensor to "High Resolution" under Rotary Motion Sensor.

8. Now click on the Channel A port to get the following menu. Select "Light Sensor" from the menu and click OK to connect that sensor. You can also click **Add sensor or Instrument** and choose Light sensor from the Analog sensors.

9.

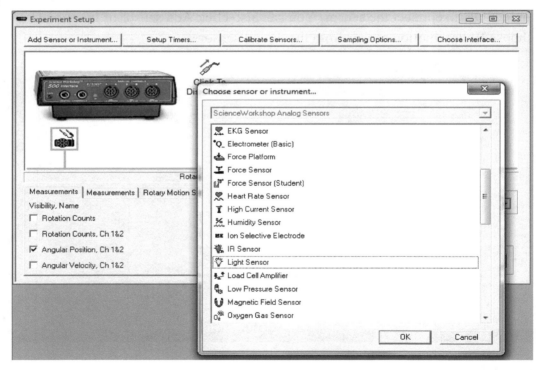

10. Double click on the Light sensor icon and set the sampling rate to 500 Hz and sensitivity, right below the sapling rate, to Medium (See the GUI below)

11. Click on the **Calculator** icon.

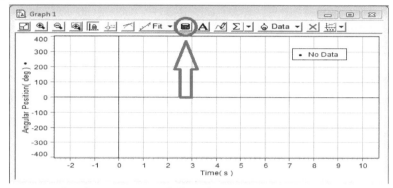

12. Upon selecting OK, the following GUI will appear.

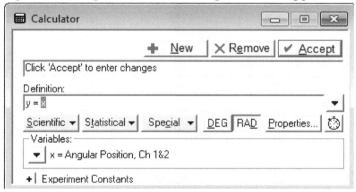

13. We now need to set the correct setting on three different focal points, the "Definition", "RAD" to "DEG", and "Variables".

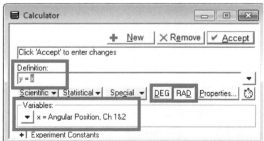

First select the "DEG" button next to the "RAD" button if not already selected. Then Change the "Variable" to x = Angular Position, Ch 1&2" if it is not already selected. Lastly, in the "Definition" box, **change y = x to y = 0.0222*x**. The click "Accept" in the top right corner and close the calculator. This will convert the angular rotation to linear units in to centimeters (cm).

14. The next step is to create graph. Click on Graph icon under Displays (3rd icon under displays).

After which you will be prompted to select a data source, select "Light Intensity Ch A(% max) and hit OK .

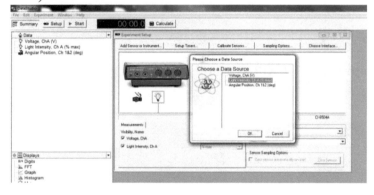

15. The Graph will appear and is now ready to be adjusted to the experiment.

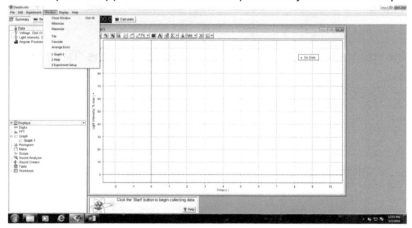

16. Now we must change the axes displayed on the graph. You will do this by clicking on the names of each axis. Make sure that the y-axis is set to "Light Intensity, Ch A (% max)", and the x-axis is set to "y = 0.0222*x".

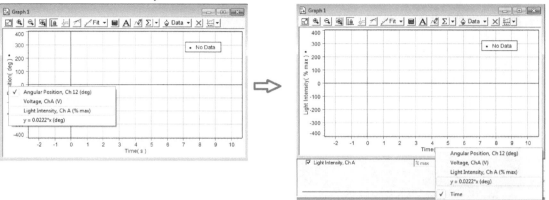

17. You will now be able to plot intensity vs position graph and measure the minima/maxima positions directly in cm and be able to use them in your calculations.
18. There are several factors that will affect your readings; use the following settings to start and change them, as needed, to get the best possible plots. For example if the peaks are chopped off, decrease the gain or decrease the aperture.
    a. Make sure the gain setting on the Light Sensor is set to 100.
    b. Double Check to see that the sampling rate is 500 Hz.
    c. Choose aperture 5 on the front of the light sensor.
19. Now that you are ready to operate the equipment, we need to make sure that the light sensor will pick up the laser pattern. This is why we left the screen board on the optical bench track. Turn on the laser and place the slit setup so that the laser passes through slit when it is perfectly vertical.

You can now remove the screen board to see where the pattern falls on the

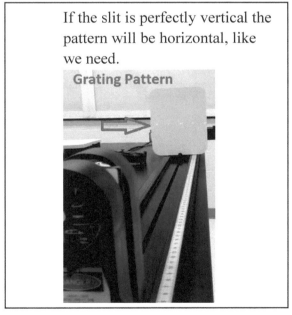

If the slit is perfectly vertical the pattern will be horizontal, like we need.

light sensor.

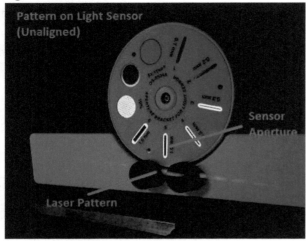

If the pattern does not fall on the sensor aperture, please call over your instructor. They will adjust the laser setting by adjusting the appropriate screws on the back of the laser. In the end your laser should fall on the sensor like so.

20. The setup is now complete and you are ready to collect your data. Follow the procedure in the next page. To plot, move the light sensor one end of the track (left or right), click **START**, and move slowly to the other end by rotating the Rotary sensor.

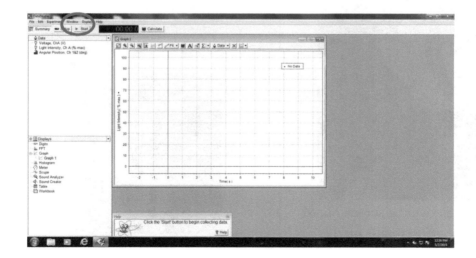

Experiment 7 and 8   Procedure (cont'd)

Data collection:

21. Center the light sensor on the track.

22. Rotate the slit disk on its mount until the laser beam falls on the 0.04 mm slit. The resulting diffraction pattern should lie in the horizontal direction, bisecting the chosen aperture.

23. Move the rotary sensor to one end of the track.

24. By turning the horizontal black rubber-edged wheel, you can move the light sensor across the diffraction/interference pattern. As you do this, the light and rotary sensors can record the intensity of the light falling on the light sensor as a function of position. Note: You do not have to be concerned with making the motion continuous because it will not affect your results. The sensor only measures the intensity of the incoming light and will not become saturated from momentary pauses as you are moving it across the track.
To start a run, Click **START** (next to the **Setup** button) and when you reach the other end of the track click **STOP** to finish the run.

25. The following steps will refine the graph:

a) The window in which the graph appears can be maximized by clicking the **select to fit** button in the upper right corner of the window.

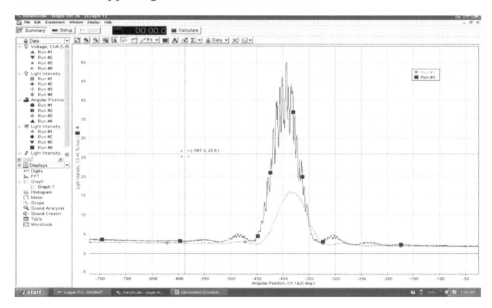

b) Zoom-in on the graph. The zoom button has a magnifying glass icon in the top left of the screen. Click this button and use the cursor to form a box containing the area you wish to zoom. Be careful not to cut off any smaller features of the plot at the ends. The button to the right of the zoom button is for "unzoom" *i.e.*, it will restore the graph to its original size.

c) It may also be helpful to remove the data points. Do this by clicking on the **Graph** icon (next to DATA) on the top and then uncheck *data points*. Once the data points are removed, the intensity profile is clearly seen as a continuous line.

26. Feel free to adjust the aperture, gain, periodic sample rate, and sensitivity to see their effect on the final plot. Use the parameters that give you the best result.

27. Print best final plot using *print* from the *file* menu. Don't forget to add your name as the identifier to pick up your plot from the printer. The graph can be improved by choosing proper parameters in the graph settings. Double click on the graph or right click on the graph to get graph settings.

28. Repeat the above procedure with other slits. Note: If you leave a graph window open, the system will plot as you move the sensor. You may try the following slits; Single slits of width 0.04 mm, and 0.08 mm, and double slits of slit width 'a' and slit separation 'd' values of (0.04, 0.25), (0.04, 0.5), (0.08, 0.25) (The numbers in parentheses refer to values in mm, in the order a, d).

Note: the choice of single slits given above has one slit larger and one smaller width, 0.04 mm and 0.08 mm, to enable you to observe the effect of changing the slit width on the separation between minima on the diffraction pattern. Similarly, the double slit combination is such that between two sets you have the same 'a' and different 'd' and another set with different 'a' and the same 'd'. (See analysis questions.) Draw the diffraction envelope by hand on the computer plot of the double slit intensity profile.

29. For your detailed data and analysis, choose one single and two double slit intensity pattern, where you see several orders of minima and maxima distinctly. Zoom-in on the graph as desired and using the x-y cursor button with cross-hairs and xy in the top (6th from the left) of the screen, next to lock icon), read out the positions and record them on your data sheets. With the single slit pattern, you will use the different order diffraction minima on either side of the center-maxima to find $(2 \cdot y_m)$. In the case of the double slit pattern, you will measure the distance between the various order interference maxima on either side of the center-maxima for determining $(2 \cdot y_m)$. You should note down positions of each of the minima (or maxima) and the corresponding order in the table of data for this experiment and then get the difference (not the difference directly). Don't forget to note down the data sampling rate, light sensor's gain, sensitivity, and the aperture used on your data sheet for each slit.

30. From your Single slit data and Eq. (D), calculate the slit width, **a** of the single slit and compare with the value specified on the slit used. Similarly, with the double slit data and Eq. (F), determine the slit-separation, **d** between the slits and compare it with the manufacture's value.

31. If time permits, you may also try other slits (e.g. continuously variable width slit and different aperture patterns) on the slit assembly.

32. Datastudio quick setup instructions:

- Double click on the Datastudio icon on the desktop. Wait for Datastudio to launch.
- Double click on Create Experiment. Datastudio window should open with a message
  "found PASCO interface 500 (or 750) and the interface is ready to use".
- Click on Add a sensor and from the Analog sensors menu choose Light sensor.
- Click on Add a sensor and from the Digital sensors menu choose Rotary sensor.

- While at the light sensor or at the rotary sensor set the sampling rate to 500Hz.
- Click on the calculator on the top and do the following in the calculator window;
- Set y = 0.0222*x, Select DEG, Variable=> Data measurement=> Angular position Channel 1&2
- Click and close the calculator
- From the left menu double click on the graph from the display
- Choose light intensity and click OK
- A graph will be displayed with Light intensity on the y-axis and time on x-axis.
- Click on Time at the bottom of the graph and choose calculator
- Now Datastudio must be ready to take data
- To check, click the start button on the top left and move the light sensor from left to right or right to left. This should draw a straight line on the graph.

Experiment 7: Single slit diffraction - Introduction

Name: Group/Section:

Partner's name: Instructor's initials:

Experiment 7: Single slit diffraction — Data

Measurement of Wavelength:

Lines per mm: _____   Lines per mm: _____

D = _____   D = _____

Lines per mm: _____

D = _____

Experiment 7: Single slit diffraction — Data (cont'd)

Measurement of Wavelength:

N	d	m	y_m	D	θ_m	λ

Wavelength, λ
(Show detailed calculations and unit conversions)

Average wavelength, λ = _____
(Show calculations)

Percent difference: _____
(Show calculations)

Experiment 7: Single slit diffraction — Data (cont'd)

Single Slit Diffraction:

Slit Width: _____   D = _____   Gain: _____

Data sampling rate: _____   Sensitivity: _____   Aperture: _____

Order No	m = 1		m = 2	
Position of the minima (Give units)	Left min	Right min.	Left min.	Right min.
Distance between m^{th} order min. ($2 \times y_m$)				
Distance from center to side (y_m)				
Calculated slit width (a)				
Percentage difference				

Show slit-width calculations and unit conversions:

Experiment 7: Single slit diffraction Analysis

Measurement of Wavelength:

(1) What happens to the spacing between the bright diffraction spots when the number of lines/mm on the grating increases? Explain with reference to your data. How will you explain your observation based on Equation A?

(2) What happens to the spacing between the diffraction spots when D is decreased? Use data taken with smaller D to answer. Explain why your measurements are better when D is large. [Hint: Think in terms of the error in the measurement of y when D is small and large.]

Single Slit Diffraction:

(1) Do the various orders give the same slit width within a reasonable error (± 5%)? If not, discuss the reasons for the discrepancy.

(2) Does the distance between diffraction minima increase or decrease when the slit width is increased? Explain with reference to your data (observation) and theory.

Summary and Conclusions

Name: Group/Section:

Partner's name: Instructor's initials:

Experiment 8: Double slit diffraction   Introduction

Experiment 8: Double slit diffraction — Data

Double Slit Interference: Trial I

Width, a = _____ Slit Separation, d = _____ D = _____

Data sampling rate: _____ Sensitivity: _____ Aperture: _____ Gain: _____

Order No	m = 1		m = 2	
Position of the maxima	Left max	Right max.	Left max.	Right max
Distance between m^{th} order maxima ($2 \times y_m$)				
Distance from center to side (y_m)				
Calculated slit separation (d)				
Percentage difference				

Show slit-separation, d calculations and unit conversions:

Experiment 8: Double slit diffraction   Data

Double Slit Interference: Trial II

Width, a = _____   Slit Separation, d = _____   D = _____

Data sampling rate: _____ Sensitivity: _____ Aperture: _____ Gain: _____

Order No	m = 1		m = 2	
Position of the maxima	Left max	Right max.	Left max.	Right max
Distance between m^{th} order maxima ($2 \times y_m$)				
Distance from center to side (y_m)				
Calculated slit separation (d)				
Percentage difference				

Show slit-separation, d calculations and unit conversions:

Experiment 8: Double slit diffraction — Analysis

Double Slit Interference:

(1) Does the distance between interference maxima increase, decrease, or stay the same when the **slit separation** is increased? Explain with reference to your data (observation) and theory.

(2) Does the distance between interference maxima increase, decrease, or stay the same when the **slit width** is increased? Explain with reference to your data (observation) and your understanding of the theory.

Summary and conclusions

Experiments 7 and 8: Review Questions

1. In an experiment the first-order maximums are measured a distance of 8 cm apart from scattering due to a diffraction grating placed 70 cm away from the screen. a) If the diffraction grating has 880 lines per centimeter what is the wavelength of light? b) If light from this laser is used in a single slit diffraction experiment at what angles would you expect to see minimums appear (please give 3 different positive angles) if the slit width is 10μm?

2 In an experiment, the diffraction pattern from a single slit was recorded with light of wavelength 700 nm on a screen held 100 cm from the slit. The distance between the first order minima on either side of the central maximum was measured to be 5.9 cm and the corresponding distance for the second order was 12.3 cm. Calculate the average width of the slit using both the first and second order data.

(3) Given the distance between the second order diffraction spots to be 24.0 cm, and the distance between the grating and the screen to be 110.0 cm; calculate the wavelength of the light. The grating had 80 lines/mm.

Answers

1. (a) 648.08 nm, 1. (b) $\theta_1 = 3.715°$, $\theta_2 = 7.446°$, $\theta_3 = 11.209°$
2. 0.023 mm   3. 678 nm

Experiments 9 and 10: Helium and Hydrogen Line Spectra

The OBJECTs of these experiments are to (i) measure the grating line spacing using Helium spectra and (ii) measure wavelengths of the visible spectral lines of hydrogen with a grating spectrograph and calculate the value of the Rydberg constant.

Theory of Atoms

In 1890 Rydberg showed that the wavelengths, λ_{jk}, of the spectral lines of atoms could be fitted by the empirical formula

$$\frac{1}{\lambda_{jk}} = Z^2 R \cdot \left(\frac{1}{j^2} - \frac{1}{k^2} \right) \quad (A)$$

where Z is the atomic number. For hydrogen, for which Z = 1, this reduces to

$$\frac{1}{\lambda_{jk}} = R \cdot \left(\frac{1}{j^2} - \frac{1}{k^2} \right) \quad (B)$$

j and k are integers chosen to fit the data, and the Rydberg constant, R, is also chosen to fit the data. It was a great achievement to be able to fit dozens of hydrogen spectral lines with a formula containing only integers and one constant R. This is a good example of a PHYSICAL LAW, a formula which summarizes large amounts of data.

Rydberg found that the four visible lines of hydrogen, the Balmer lines, could be fitted by assigning j = 2 and k = 3, 4, 5, 6. Given a measurement of the red Balmer line, he found that its wavelength should be characterized by j = 2 and k = 3 and R could be calculated from equation (B) as:

$$R = \frac{1}{\lambda_{jk}} \left(\frac{1}{j^2} - \frac{1}{k^2} \right)^{-1} \quad (C1)$$

For the red Balmer line: $R = \frac{1}{\lambda_{23}} \left(\frac{1}{2^2} - \frac{1}{3^2} \right)^{-1}$ where λ_{23} is the wavelength. (C2)

The same value of R resulted when he assigned k = 4, 5, 6 to the shorter wavelength lines and did a similar calculation. The best value of R is now known to be

$$R = 1.096776 \times 10^7 \text{ m}^{-1} = 1.096776 \times 10^{-2} \text{ nm}^{-1} \quad (1 \text{ nm} = 10^{-9} \text{m})$$

We normally measure the wavelengths λ in nanometers (nm), and since R has the units of $(1/\lambda)$, R has the units 1/nm.

Experiments 9 and 10: Helium and Hydrogen Line Spectra

In the "normal" course of science, one first has data, followed by a law summarizing the data, and then a general theory that explains the law and many other phenomena besides. Science does not always progress this way. It did in the case of atomic spectra. Niels Bohr in 1913 not only explained the general form of the Rydberg formula, he also predicted precisely the value of the constant R, and he predicted many new atomic phenomena. Bohr was guided by Rutherford's "solar system" model of atoms (1910) and by Planck's proposed quanta (bundles) of electromagnetic energy E = hf.

E = energy in joules, f = frequency in Hz, and h = 6.6 x 10^{-34} joule sec = Planck's constant, a very small but important constant. Rutherford had proposed that an atom consisted of negatively charged, very light electrons circling a massive, positively charged nucleus and that the electric force of attraction held the electrons in orbit. According to classical theory, constantly circling electrons should lose energy by radiating electromagnetic waves (light), and hence all atoms should collapse, and we would not exist. But we do exist! Bohr's cure for this contradiction was beautifully simple. (a) Classical theory does not apply to the very small orbits of atomic electrons; electrons in stable orbits do not radiate and lose energy. He then proposed (b) that electrons do radiate when they fall spontaneously from a higher (E_k) to a lower (E_j) energy orbit and that the energy radiated would be E = $E_k - E_j$. The electron must first have been excited by collision to the higher energy orbit. In order to explain the fact that spectral lines occurred at only certain discrete wavelengths, he further proposed that (c) only certain stable electron orbits are allowed. He proposed the strictly ad hoc rule that only orbits with angular momenta L = nh/2π would be allowed, (n is any integer except 0.). He then used his model to derive the Rydberg formula and found that the integers j and k corresponded to the integer n for the lower and higher energy orbit, respectively. He predicted R to be

$$R = \frac{me^4}{8\varepsilon_0^2 h^3 c} \quad (D)$$

where m is the mass of the electron, e is the charge of the electron, c is the speed of light, h is the Planck's constant, and $1/4\pi\varepsilon_0 = k_e$ is the constant in Coulomb's electric force law.

Note that equation (D) contains quantities which are very small, $e^4 = (1.6 \cdot 10^{-19})^4$ (coulomb)^4 and very large, c = 3 x 10^8 m/s. The fact that Bohr predicted R to a fraction of a percent was a major triumph.

Experiments 9 and 10: Helium and Hydrogen Line Spectra

Spectroscopes

A spectroscope is a device in which a prism, or a diffraction grating, is used to separate light into its component spectral lines. If you view the lines directly, the device is called a "spectroscope"; if the component lines are recorded on photographic film (or other detecting device), it is called a "spectrograph" or a "spectrometer".

In Fig. 1 below, we see a light source R placed close to the entrance slit S_1 of a spectroscope. Light diverging from the slit falls on a converging lens L_1.

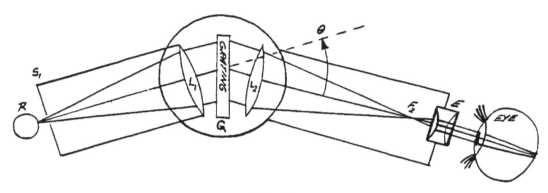

Fig. 1

The lens-slit distance is adjusted to equal the focal length of L_1 so that parallel rays will emerge from L_1 and strike the diffraction grating G. S_1 - L_1 is called the "collimator". A "telescope" L_2 – F_2 – E focuses only parallel rays leaving the grating on the cross hairs positioned at F_2. (Note that the cross hairs themselves are not shown in Fig. 1.) The telescope is mounted so that its angle θ can be precisely read. The distance L_2 - F_2 is adjusted to equal the focal length of the converging lens L_2 so that an image of the slit S_1 falls exactly at the cross hairs at F_2. The image is viewed with a magnifier E (eyepiece), which is positioned so that it sends parallel rays into the eye. The eye then forms an image on its retina and you see the slit image. In a spectrograph, a photographic film is placed at F_2.

A diffraction grating is a plate on which very closely spaced parallel lines are ruled. Our gratings are glass plates on which a thin plastic film is mounted. The plastic film is a replica of a very precisely ruled metal grating. The separation d of the lines must be larger than the wavelength of visible light, but small enough so that hundreds of them are illuminated by the light from the source.

If we assume d = 2000 nm = 2 μm = 2×10^{-6} m, we see that the number N of lines per meter is N = 500,000 lines/m = 5,000 lines/cm. 5,000 lines/cm is typical for a grating you will use.

Experiments 9 and 10: Helium and Hydrogen Line Spectra

When light of wavelength λ falls on a grating, the emerging light is strong only at the angles θ_m for which the path difference between rays from adjacent slits is an integer number of wavelengths mλ. (Fig. 2). Recall the theory of diffraction from experiment 7.

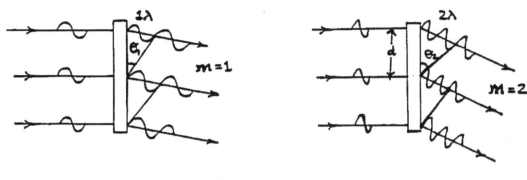

diffraction

Fig. 2

When m = 1, 2, 3 . . . we get the first, second, third . . . order spectral lines of that color. Given light of two or more different wavelengths, we get sets of lines corresponding to m = 1, 2, 3 . . . with red lines at larger angles than blue lines for each m as θ depends directly on λ. The grating formula is

$$m\lambda = d \cdot \sin\theta_m \qquad (E)$$

The grating you will use will only give the first and second order (m = 1, 2) spectra. The light seen at θ = 0° will be the color of the source, as none of the lines will be separated in the straight-ahead light.

In this experiment, you will use the 587.57 nm yellow line of helium as a wavelength standard. First you will measure the angles θ of the helium yellow line (first and second order), which will give two determinations of the grating line spacing d (see equation (E)). The measurements of the angles of the first and second order red, blue-green, and violet hydrogen lines and the knowledge of d will allow you to calculate the wavelengths of these three hydrogen lines. From these three wavelengths we can get three determinations of the Rydberg constant R using j = 2 and k = 3, 4, 5 for the red, blue-green, and violet lines, respectively. See equation (C).

EQUIPMENT
1 - spectroscope, mounted diffraction grating, and support for black cloth
1 - spectrum tube power supply
1 - helium spectrum tube
1 - hydrogen spectrum tube
1 - black cloth
2 - plastic, hand-held diffraction grating
for the room: 2 extra helium and 2 extra hydrogen spectrum tubes
        2 – blue plastic mounts to adjust the plane of grating

SAFETY PRECAUTION
BEFORE YOU CHANGE SPECTRUM TUBES, SWITCH THE TUBES OFF AND UNPLUG.

Experiments 9 and 10: Helium and Hydrogen Line Spectra

BEFORE YOU BEGIN, the instructor will do the following.
(a) Show you how to adjust the position of the spectrum tube.
(b) Show you how to adjust the eyepiece focus, the telescope focus, the telescope angle clamping screw, and fine adjust (tangent) screw.
(c) Demonstrate the idea of parallax and explain its use in focusing the telescope.
(d) Urge you to manipulate the spectroscope controls gently.
(e) Show you how to use the black cloth and the flashlight when reading the weaker lines.
(f) Have you look at a helium and a hydrogen discharge tube with a hand-held diffraction grating so that you can see all of the spectral lines you will be measuring one at a time with the spectroscope.

Measuring Angles

The angles θ of the spectral lines are the averages of the left and right angles (Fig. 3).

$$\theta(L) = [A(L) - A(0)] \text{ and } \theta(R) = [A(0) - A(R)] \tag{F}$$

$$\theta_{av} = \frac{1}{2}[\theta(L) + \theta(R)] = \frac{1}{2}[A(L) - A(R)]$$

A(L) and A(R) are the angular positions of the left and right spectral lines measured on the spectroscope. A(0) is the angular position of the center slit image. The left and right angles should differ by less than about 6' (or 0.1°). For a given spectral line, the ratio $\sin\theta_2 / \sin\theta_1$ of the sines of the second and first order angles should be within about 2% of 2.00. This serves as another check on your readings.

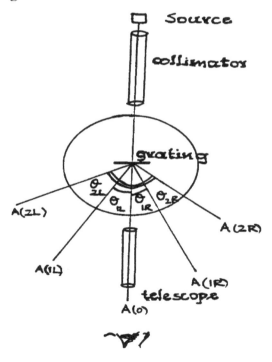

Fig. 3

Experiments 9 and 10: Helium and Hydrogen Line Spectra

Reading the spectroscope vernier angle scale: Use only the scale window to your right as you face the light source. [We will not take advantage of the slightly improved precision gained by averaging the readings from the left and right windows.]

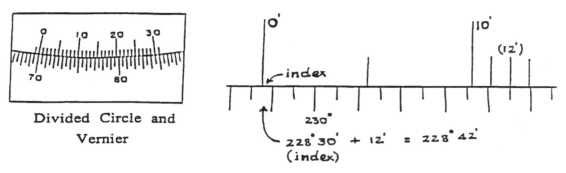

Fig. 4

The outer scale divisions are marked every 0.5° (= 30' of arc) and are the angle scale. The inner vernier scale divisions have about the same separations and are labeled from -1' to +31' of arc. The 0' vernier line is the index for reading the approximate angle on the outer scale. In Fig. 4 (left part), the 0' index line is exactly aligned with the 70° scale mark. Note that other vernier lines, which are closer together by a factor of 29/30 than the scale marks, fall progressively to the left of the scale line below. The 30' (vernier) line is again directly aligned with a scale mark. The angle is 70° 00' since we have 70° 00' (index) + 00' (vernier) or 69° 30' (index) + 30' (vernier). The right part of Fig. 4 shows a setting of 228° 42', as we have an index position beyond 228° 30' and alignment of the 12' vernier. Had the 0' index been to the right of the 228° 00' scale mark the resultant angle would have been 228° 12'. In sum, the vernier scale gives us a precise interpolation between the 30' marks on the angle scale.

There are three steps in reading angles. (a) Identify the outer scale mark just to the left of the vernier 0' index mark. (b) Locate the vernier scale mark that is aligned with an outer scale mark. (c) Add the vernier minutes to the outer scale mark degrees and minutes.

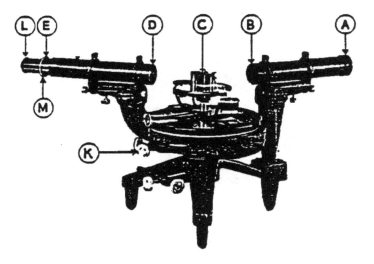

A - Slit
B - Collimator objective
C - Dispersing prism
D - Telescope objective
E - Position of cross hairs
L - Eyepiece Ring
M - Telescope focusing ring
K - Telescope arm clamping and tangent screws

Fig. 5

Experiments 9 and 10: Helium and Hydrogen Line Spectra

<u>Telescope angle adjustment:</u> (Fig. 5) Loosen telescope angle clamping screw (K) in order to swing the telescope freely. After the clamping screw is gently tightened, the telescope may be moved precisely with the tangent screw (K).

Practice Reading Angles. RECORD the readings to the nearest minute of both you and your partner at two different angular positions of the telescope. If your readings agree to within 2', you are ready to take data.

<u>Position the helium spectrum tube:</u> With the HIGH VOLTAGE OFF AND UNPLUGGED, position the midsection of the He spectrum tube about 1/2 mm from the end of the collimator and center it laterally. Swing the telescope to one side, turn on the high voltage, and view the entrance slit directly through the collimator. Move the spectroscope slightly to the left and right and find the position for which the slit illumination is brightest.

<u>Focusing the telescope cross hairs:</u> Set the telescope straight ahead and bring the slit image to the center of the field of view. Move the eyepiece (L in Fig. 5) in and out to get the sharpest image of the cross hairs. Get a focus without your glasses if possible, as the field of view is wider with your eye close to the eyepiece. If you can't get a good focus without your glasses, leave them on and get a focus. Each partner should readjust for best focus. The eyepiece focus may be readjusted at any time without changing the alignment of the spectroscope or changing the readings.

<u>Focusing the slit image without parallax error:</u> Set the telescope angle so that the cross hairs intersect in the middle of the slit image. Twist the knurled chrome ring on the telescope (E in Fig. 5.) to get the sharpest slit image focus. Then move your eye slightly left and right and look for any horizontal motion of the slit relative to the crosshairs (parallax). Adjust the knurled ring to get an opposite relative motion and then readjust for no motion (no parallax). This is an extremely important adjustment, as you do not want the adjustment of the telescope angle to depend on the exact location of your eyeball.

<u>Setting the telescope position:</u> Always position the intersection of the cross hairs on the right edge of each left <u>and</u> right spectral line and the central slit image. See Fig. 6. A slit edge may be positioned more precisely than the slit center, and the resulting angular displacements θ are the same as if the center positions had been used.

Fig. 6

Experiments 9 and 10: Helium and Hydrogen Line Spectra

General comment for measurements and calculations: **This experiment allows such an accuracy that all intermediate values should be kept to at least <u>five</u> significant digits, if <u>all</u> the adjustments described on the previous page have been done correctly.**

HELIUM YELLOW LINE ANGLES.

Each partner should take a set of angular position readings for the center slit image A(0), and for the left and right first and second order yellow helium line, A(1L), A(1R), A(2L), A(2R). Record the positions in degrees and minutes as you read them on the spectroscope and convert these to decimal degrees before you do your calculations. Both partners should record both sets of data.

CALCULATE the left and right angles for both sets of data using Equation F. They should differ by less than 0.1°.

CALCULATE the average first and second order angles and find $\sin\theta_2 / \sin\theta_1$. Your value of $\sin\theta_2 / \sin\theta_1$ should be within 1% of the expected value 2. Therefore, you should keep the angles to at least two decimal places and report $\sin\theta_2 / \sin\theta_1$ to three decimal places.
Call the instructor if you appear to have scale reading, or spectroscope alignment problems.

HYDROGEN LINE ANGLES.

Unplug the high voltage spectrum tube supply, and replace the helium with the hydrogen spectrum tube.

RECORD one set of readings by one partner of the angular positions of the first and second order red, blue-green, and violet hydrogen lines (The second order violet line may be too weak to measure in which case you will measure only the first order violet line.)

CALCULATE $\theta(L)$, $\theta(R)$, and $\theta(av)$ for each line. Use that partner's A(0) from step the helium yellow line angles to calculate these.

CALCULATE $\sin\theta_2 / \sin\theta_1$ for the red, blue-green, and violet lines.
Review your results to make sure you have correct readings.

Name: Group/Section:

Partner's name: Instructor's initials:

Experiment 9: Helium Line Spectra Introduction

Experiment 9: Helium Line Spectra — Data

Angle: _____

Partner's Angle: _____   Estimate of error, $\Delta\theta$ = _____
(This is the reading error obtained from difference between yours and your partner's angle)

Helium Yellow Line Angles:

$A(0)$ = _____

First order lines:	Second order lines:
$A(1R)$ = _____	$A(2R)$ = _____
$A(1L)$ = _____	$A(2L)$ = _____
$\theta(1R)$ = _____	$\theta(2R)$ = _____
$\theta(1L)$ = _____	$\theta(2L)$ = _____
θ_1 = _____	θ_2 = _____

$$\frac{\sin\theta_2}{\sin\theta_1} = \text{_____}$$

Show all calculations:

Percent Difference: _____

Experiment 9: Helium Line Spectra   Analysis

GRATING LINE SPACING. Use the helium data to find the average values of $\theta(1)$ and $\theta(2)$ for the first and second order helium 587.57 nm yellow line. Using these values calculate two values of the grating spacing $d(1)$ and $d(2)$ in nanometers. Average these and calculate the number of lines/cm of the grating, $N = 1/d(ave)$.
(Show calculations and unit conversions)

Summary and Conclusions

Name: Group/Section:

Partner's name: Instructor's initials:

Experiment 10: Hydrogen Line Spectra Introduction

Experiment 10: Hydrogen Line Spectra   Data

Hydrogen Line Angles

A(0) = _____

First Order Lines:

Red:

A(1L) = _____

A(1R) = _____

θ(L) = _____

θ(R) = _____

θ(1av) = _____

Blue-Green:

A(1L) = _____

A(1R) = _____

θ(L) = _____

θ(R) = _____

θ(1av) = _____

Violet:

A(1L) = _____

A(1R) = _____

θ(L) = _____

θ(R) = _____

θ(1av) = _____

Experiment 10: Hydrogen Line Spectra   Data

Hydrogen Line Angles

Second Order Lines:

Red:

A(2L) = _____

A(2R) = _____

θ(2L) = _____

θ(2R) = _____

θ(2av) = _____

Blue-Green:

A(2L) = _____

A(2R) = _____

θ(2L) = _____

θ(2R) = _____

θ(2av) = _____

Violet:

A(2L) = _____

A(2R) = _____

θ(L) = _____

θ(R) = _____

θ(2av) = _____

Angle Checks:

Red: $\sin\theta_2 / \sin\theta_1 =$ _____

Percent Difference: _____

Blue-Green: $\sin\theta_2 / \sin\theta_1 =$ _____

Percent Difference: _____

Violet: $\sin\theta_2 / \sin\theta_1 =$ _____

Percent Difference: _____

Experiment 10: Hydrogen Line Spectra   Analysis

1. **HYDROGEN WAVELENGTHS**. From the experiment 9, use the value of grating spacing d(1) in Equ. (E) to calculate the first order hydrogen wavelengths for red, blue-green, and violet.
   Then use d(2) and the second order angles to calculate the red, blue-green, and violet wavelengths.
   Average the wavelength from first and second order measurements for these three hydrogen lines.
   Important: If your d spacing from experiment 9 has more than 5% error, don't use your value. Use the standard d spacing for the grating you are using.

2. **RYDBERG CONSTANT**. Calculate three values of the Rydberg constant R using the average hydrogen wavelength of each color in Question (1).
   In equation (C), j = 2 for all lines; k = 3, 4, 5 for the red, blue-green, violet lines, respectively.
   Therefore, λ_{red} (ave) = λ_{23}; λ_{bg} (ave) = λ_{24}; λ_{violet} (ave) = λ_{25};
   Average the three values of R.
   What is R(av) ± ΔR(av)?
   Find the % difference between R(av) and the accepted value.

Experiment 10: Hydrogen Line Spectra — Analysis

(3) ERRORS
 (a) How well do you think you could set the spectroscope and read the angles in this experiment?
 (b) Does your estimated $\Delta\theta$ explain the variations seen in the wavelength and R calculations? Change the angle by 0.2° and recalculate the wavelength and R for one of your values. Comment on the changes you find.

Summary and Conclusions

Experiments 9 and 10: Review Questions

1. A grating has 6250 lines/cm. What is the grating spacing in nm?

2. For the above grating calculate the angle of a second order orange line of wavelength 600 nm.

3. Convert the spectroscope reading 228°16′ to decimal degree reading.

4. In an experiment you measure a first-order red line for Hydrogen at an angle difference of $\Delta\Theta$ = 22.78°. The diffraction grating you are using has 5900 lines per cm.
a) What is the wavelength of this light?
b) What is the value of Rydberg's constant for this measurement?

Answers:

    1. 1600 nm.   2. 48.59°   3. 228.266°   4. (a) 656.3 nm and (b) $R = 1.097 \times 10^7 \text{ m}^{-1}$

Experiments 11 and 12: Nuclear Counting Statistics and Interaction of radiation with matter

Counting Statistics

The random uncertainty in most physical measurements is due to a number of small, unknown physical and physiological sources of random variation. Hence it can only be found by observing the scatter in repeated trials. Random uncertainty is usually given as a "standard deviation" which is defined as the square root of the average of the squared deviations.

$$\sigma = \sqrt{\frac{\Sigma(x_i - \bar{x})^2}{(N-1)}} \qquad (A)$$

If the factor(s) causing the variation in x are truly random, then 32% of the trials x will fall outside $\bar{x} \pm \sigma$ i.e., outside one standard deviation from the mean. About 5% should fall outside of $\bar{x} \pm 2\sigma$ i.e., outside two standard deviations from the mean. The standard deviation σ in the number of events C resulting from a single, spontaneous physical process can be predicted after just one measurement and will be

$$\sigma = \sqrt{C} \qquad (B)$$

This equation is obviously only valid when C and σ are dimensionless. If we make N repeated trials for the number of events C, then the mean value, $\bar{C}$, of the N trials is better information about σ and

$$\sigma = \sqrt{\bar{C}} \qquad (C)$$
(better value of σ when N trials are taken)

σ is the random uncertainty of a *single trial*. The *standard deviation*, σ_N, *of the mean*, $\bar{C}$ is a factor of $\sqrt{1/N}$ less than the *standard deviation*, σ of a single trial.

$$\sigma_N = \frac{\sigma}{\sqrt{N}} = \sqrt{\frac{\bar{C}}{N}} \qquad (D)$$
(standard deviation of the mean $\bar{C}$)

The spontaneous decays of atoms by light emission and the spontaneous decays of nuclei by particle or gamma ray emission are physical events controlled by a single random process and the standard deviations predicted by equations (B) and (C) apply. Equation (D) applies to random errors of all kinds.

In this experiment, we will take **five** sets of N = 10 repeated readings of the number of emissions, or counts C in a time interval Δt = 10 s of beta rays (fast electrons) from a source of long-lived radioactive nuclei whose average decay rate is constant during the laboratory period. By changing the distance between the radioactive source and the detector, we can get average counts $\bar{C}$ of tens, hundreds, and thousands in 10 sec. For each of the five sets of data, we will calculate $\bar{C}$ and σ. We will then look to see how many trials fall outside of $\bar{C} \pm \sigma$. While 3 of 10 trials (32%) are expected to fall outside one standard deviation, the number $3 \pm \sqrt{3} = 3 \pm 1.7$ is itself statistical. If we look at all 30 trials, then the total events falling outside one standard deviation should be $9 \pm \sqrt{9} = 9 \pm 3$. i.e., 32% of the time! For example, suppose you find that 6 of your 30 trials fall outside of one standard deviation. This is quite normal. However, if none of the 10 trials fall outside of the range $\bar{C} \pm \sigma$, then something may be wrong with your equipment.

Experiments 11 and 12: Nuclear Counting Statistics and Interaction of radiation with matter

Interaction of Radiation with Matter - Attenuation of Gamma-particles

Gamma-rays are photons (high energy "light") and it must interact with materials to be seen (or detected). Most of the radioactive materials emit γ-particles. The Cs-137 button source used in our experiments emits γ particles with an energy of 616 keV. This experiment is on the interaction of γ-particles with matter.

Gamma rays were first identified in 1900 by Becquerel as a component of the radiation from uranium and radium that had much higher penetrability than alpha and beta particles. In 1909, Soddy and Russell found that gamma-ray attenuation followed an exponential law and that the ratio of the attenuation coefficient to the density of the attenuating material was nearly constant for all materials.

The attenuation of Gamma-Rays

Figure 1 illustrates a simple attenuation experiment. When gamma radiation of intensity I_0 is incident on an absorber of thickness t, the intensity I of the transmitted radiation is given by

$$I = I_0 \exp(-\alpha t), \qquad (E)$$

where t is the thickness of the material and □ is the linear attenuation constant (expressed in units of cm^{-1}). The ratio I/I_0 is called the gamma-ray transmission. The transmission increases with increasing gamma-ray energy and decreases with increasing absorber thickness.

Similarly, if N_0 is the number of incident gamma particles and N is the number of outgoing gamma particles, then

$$N = N_0 \exp(-\alpha t). \qquad (F)$$

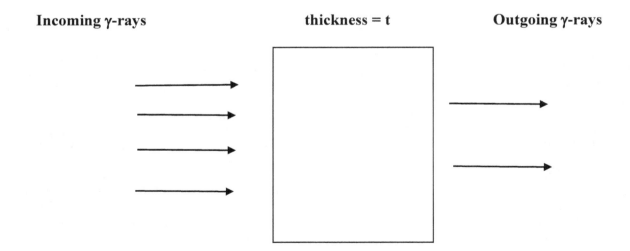

Experiment 11: Nuclear Counting Statistics   Procedure

The OBJECTS of the experiment are:
(1)   to test the prediction that the standard deviation of nuclear counts C will be $\sigma = \sqrt{C_{net}}$;

BEFORE YOU BEGIN, the instructor will:
(a)   show you how to place the source under the support shelf;
(b)   show you how to adjust and operate the timer and counter;
(c)   review the following:

PRECAUTIONS:

I.   Keep your sealed radioactive source on the side table when it is not in use. (Sealed microcurie sources are perfectly safe, but nuclear radiation, however small, should be minimized. Millicurie sources are 1000 times stronger and do present safety concerns. One CURIE equals 3.7×10^{10} counts/second.)

II.   The Geiger counter has a thin end window that can be broken. Keep objects away from the end of the Geiger counter.

EQUIPMENT

1 - Geiger counter and assembly
1 - interval timer
1 - Cs 137 sources sealed in plastic kept in a plastic box

Procedure for ST 360 - GM counter setup and counting instructions

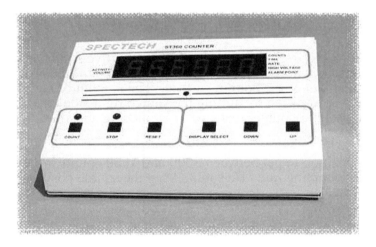

Setting it up

We use right 3 buttons to setup the instrument and left 3 buttons to count.

Press **DISPLY** once: The red LED in the top right corner of the display will light up next to one of these;

Counts, Time, Rate, High voltage or **Alarm point**.

Press **DISPLY** repeatedly until **High voltage** is chosen (Red LED next to it lit).
Press **UP/DOWN** buttons to set the high voltage. Set the voltage to a value given by your instructor.
Press **DISPLY** repeatedly until **TIME** is chosen (Red LED next to it lit).
Press **UP/DOWN** buttons to set the time to the desired value (10s or 100s). When you set time to 10s, and press COUNT, the unit will record counts for 10s and stops automatically.
Press **DISPLY** repeatedly until **COUNTS** is chosen (Red LED next to it lit). Unit is ready to take data.

Taking data
Place the Cs 137 source in the GM tube stand using the plastic tray.
Press the **RESET** button from the left 3 button group. Display should read all zeros.
Press the **COUNT** button, the unit will start taking data. It will stop automatically at the end of preset time set during the setup.
Use the **STOP** button, if you want to terminate counting in the middle of counting and restart.
If you need to increase or decrease the counting time, use the procedure described in the setup part to do so. **Don't forget to return the unit to display**.
If you need to increase or decrease the high voltage, use the procedure described in the setup part to do so. Don't forget to return the unit to display

Experimental details
Use the above procedure to take data and record it on page 142. Change the sample position with respect to the GM tube to adjust number of counts in each case. If counts are high move the source to lower racks on the stand. If low, move it up. If you reached bottom of the stand and still the counts are high, **flip the source or move it slightly to the front**.
We will not use Rate and Alarm modes.

Counting Statistics Data and Analysis:

(1) <u>BACKGROUND COUNTS</u>. Remove the Cs137 source and record the numbers of background counts in a 100 s run. Background counts are due to cosmic rays and the radioactivity of building materials. If the number is higher than 50 in 100 s, ask the instructor to check your equipment.

(2) <u>10 SECOND COUNTING RUNS</u>. Place the Cs137 radioactive source is near the top surface of the plastic disc, so more nuclear radiations come out of the top (lettered) surface of the source than out of the bottom surface. The amount of radiation reaching the Geiger counter may be adjusted by turning the source right side up, or up side down, and/or by changing the source-counter distance or by placing absorbers between the source and the detector.

   (a) Adjust the source position to get between 3000 and 4000 counts in 10 s. RECORD ten 10 s runs at this counting rate. Be sure to keep the source position the same during the ten runs.
   (b) Adjust the source position to get between 1000 and 2000 counts in 10 s. RECORD ten 10 s runs at this counting rate.
   (c) Adjust the source position to get between 500 and 700 counts in 10 s. RECORD ten 10 s runs at this counting rate.
   (d) Adjust the source position to get between 100 and 200 counts in 10 s. RECORD ten 10 s runs at this counting rate
   (d) Adjust the source position and orientation to get between 20 and 40 counts in 10 s. RECORD ten 10 s runs at this counting rate.

(4) CALCULATE THE AVERAGE $\bar{C}$ and the PREDICTED STANDARD DEVIATION σ ($=\sqrt{\bar{C}}$) for each of your five sets of N = 10 runs. Circle the runs in each of the five sets that fall outside of $\bar{C}\pm\sigma$. Compare your results with the predictions that 3 of the 10 runs and that 15 of the 50 runs will fall outside of one standard deviation from the mean. Since these are statistical predictions, the results often fall between 8 and 20 out of 30 runs, and occasionally vary from the prediction even more. Does your equipment appear to be working satisfactorily? Comment briefly.

(5) CALCULATE THE RELATIVE ERROR $\sigma/\bar{C}$ for each of your five sets of data.

Does the <u>error</u> σ get larger, or smaller, as C increases?
Does the <u>relative error</u> $\sigma/\bar{C}$ get larger, or smaller, as $\bar{C}$ increases?

Experiment 12: Attenuation of Gamma Particles — Procedure

Attenuation of Gamma-rays by Aluminum and Mylar:
In this experiment, you will calculate the linear attenuation constant α for Aluminum and Mylar.

Procedure:
1. Place the Cs-137 source about 3 cm below the detector.
2. Find the number of counts for 10 s. You should get ~ 1100 counts for 1 micro-Curie (μCi) sources and ~ 6000 for 5 μCi sources.
3. Repeat 2 more times. Find the average count (N_0, no absorber between the source and the detector).

Aluminum:

4. Cut the Al-foil into 1″ x 2″ strips. You may need 10-14 strips. The instructor will let you know the thickness of the Al-foil.
5. Place 2 strips on the source. Find the counts for 10 sec. Repeat 2 more times. Find the average of the 3 trials.
6. Add 2 more strips for a total of 4. Repeat step 5.
7. Find the counts for a total of 10-12 strips.
8. Draw a graph of N versus total <u>thickness of Aluminum t (in cm)</u>. This graph will show exponential decay (similar to capacitor discharge).
9. Draw a graph of ln (N/N_0) versus t. This will be a straight line. <u>Find the slope from linear fit to the data</u>.
10. The slope is the linear attenuation constant α (in units of cm^{-1}).

Mylar

Repeat above steps (1-10) for Mylar. Find α from data analysis.

Experiment 11: Nuclear Counting Statistics

Name: Group/Section:

Partner's Name: Instructor's initials:

Experiments 11: Nuclear Counting Statistics - Data

Counting Statistics Data

(1) High voltage Setting: _____

(2) Background Counts: _____

(3) Circle values outside of $\bar{C} \pm \sigma$.

Counting range:_____

Trial	Counts
1	
2	
3	
4	
5	
6	
7	
8	
9	
10	

Counting range:_____

Trial	Counts
1	
2	
3	
4	
5	
6	
7	
8	
9	
10	

Counting range:_____

Trial	Counts
1	
2	
3	
4	
5	
6	
7	
8	
9	
10	

Counting range:_____

Trial	Counts
1	
2	
3	
4	
5	
6	
7	
8	
9	
10	

Trial	Counts
1	
2	
3	
4	
5	
6	
7	
8	
9	
10	

Experiments 11: Nuclear Counting Statistics - Data

(4)

$\bar{C} =$ _____ $\bar{C} =$ _____ $\bar{C} =$ _____

$\sigma =$ _____ $\sigma =$ _____ $\sigma =$ _____

$\dfrac{\sigma}{\bar{C}} =$ _____ $\dfrac{\sigma}{\bar{C}} =$ _____ $\dfrac{\sigma}{\bar{C}} =$ _____

$\bar{C} \pm \sigma =$ _____ $\bar{C} \pm \sigma =$ _____ $\bar{C} \pm \sigma =$ _____

$\bar{C} =$ _____ $\bar{C} =$ _____

$\sigma =$ _____ $\sigma =$ _____

$\dfrac{\sigma}{\bar{C}} =$ _____ $\dfrac{\sigma}{\bar{C}} =$ _____

$\bar{C} \pm \sigma =$ _____ $\bar{C} \pm \sigma =$ _____

(5)   The absolute error σ : _____ As $\bar{C}$ increases

   The relative error $\dfrac{\sigma}{\bar{C}}$: _____ As $\bar{C}$ increases

Experiments 11: Nuclear Counting Statistics - Analysis

(1) Refer to the table from step (4) in answering the following questions.
    (a) How does the <u>absolute</u> error in C_{net} vary with C_{net}?
    (b) How does the <u>relative</u> error in C_{net} vary with C_{net}?
    (3) What do your answers to (a) and (b) imply about the precision of the data measured in this experiment?

Summary and conclusion

Experiment 12: Attenuation of Gamma Particles - Introduction

Name: Group/Section:

Partner's Name: Instructor's initials:

Experiments 12: Attenuation of Gamma Particles - Data

Aluminum
Thickness of each layer = _____ Density of Al = 2.7 g/cm^3

Cs-137:   1 μCi   or   5 μCi   (circle the source strength you are using)

Approximate Source to detector distance:

Number of Layers	Thickness t (cm)	Counts for 10 sec. N_1	N_2	N_3	N(average)	ln (N/N$_0$)
0					N_0 =	
2						
4						
6						
8						
10						
12						

Draw graphs of: (i) N vs t   (ii) ln (N/N$_0$) vs t

Linear attenuation constant, α_{Al} =
(Slope of ln (N/N$_0$) vs t)

α_{Al} / d_{Al} =

Mylar

Thickness of each layer = _____   Density of Mylar = 1.38 g/cm³

Cs-137:   1 μCi   or   5 μCi   (circle the source strength you are using)

Approximate Source to detector distance:

Number of Layers	Thickness t (cm)	Counts for 10 sec. N_1	N_2	N_3	N(average)	ln (N/N$_0$)
0					N_0 =	
2						
4						
6						
8						
10						
12						

Draw graphs of: ln (N/N$_0$) vs t

Linear attenuation constant: α_{Mylar} =
(Slope of ln (N/N$_0$) vs t)

α_{Myl} /d_{Myl} =

Analysis
(1) Compare the values of: α_{Al}/d_{Al} and α_{my}/d_{my}. Are they same? Comment on your result.

Summary and Conclusions: